ELSEVIER'S DICTIONARY OF TOOLS AND IRONWARE

ELSEVIER'S DICTIONARY OF TOOLS AND IRONWARE

IN SIX LANGUAGES

ENGLISH/AMERICAN, FRENCH, SPANISH, ITALIAN, DUTCH AND GERMAN

compiled by

W.E. CLASON

Zoutelande, The Netherlands

ELSEVIER SCIENTIFIC PUBLISHING COMPANY
AMSTERDAM – OXFORD – NEW YORK 1982

ELSEVIER SCIENTIFIC PUBLISHING COMPANY
Molenwerf 1
P.O. Box 211, 1000 AE Amsterdam, The Netherlands

Distribution for the United States and Canada:

ELSEVIER SCIENCE PUBLISHING COMPANY INC.
52, Vanderbilt Avenue
New York, N.Y. 10017

Library of Congress Cataloging in Publication Data

Clason, W. E., 1898-
Elsevier's dictionary of tools and ironware.

1. Tools--Dictionaries--Polyglot. 2. Hardware--Dictionaries--Polyglot. 3. Dictionaries, Polyglot.
I. Elsevier Scientific Publishing Company. II. Title.
TJ1195.C53 621.9'003 82-5108
ISBN 0-444-42085-1 AACR2

ISBN 0-444-42085-1

Electronic data processing:
Büro für Satztechnik W. Meyer KG
Hanau, W. Germany

Printed in The Netherlands

PREFACE

A six-language dictionary of tools and ironware covers two fields that are closely linked and so wide-ranging in scope that one is obliged to be rather selective in order to keep the size of the work within reasonable bounds.

I am conscious that this may leave me open to criticism in that too little or too much has been included from one sector or another. But I feel sure that these imperfections have been kept to a minimum owing to the assistance of the various international organisations and the use of modern catalogues from the Netherlands, Britain, France, Spain, Italy and Germany.

In compiling this dictionary I have been guided not by scientific nomenclature but rather by the technical language of this field.

For my wife (my faithful collaborator) and myself this will be our last dictionary. Our sincere thanks go to those who have assisted us in its preparation.

W.E. Clason

ABBREVIATIONS

agr	agriculture
art	artist's tools
axe	axes
ben	bending tools
bre	brewery
bui	building
bur	burners
car	carpenting
che	chemistry
chi	chisels and gouges
civ	civil engineering
cla	clamping
cle	cleaning
con	control
coo	cooperage
cov	covering
cpt	component parts
cut	cutting
dec	decoration
dra	drawing
dre	dredging
dri	drilling
ele	electricity
eng	engineering
far	files and rasps
fas	fastenings
fin	finishing
fir	fire brigade
for	forging
fou	foundry
foy	forestry
fre	cutting/milling
gas	gold- and silversmiths
gen	general
gla	glass working
gri	grinding
ham	hammers
hor	horticulture
hou	household appliances
hyd	hydraulics
iwa	ironware
joi	joining
lat	lathes
lea	leather working
lub	lubrication
mas	masonry
mea	measuring apparatus
mec	mechanics
met	metal working
min	mining
nav	navigation
nut	nutrition
off	office equipment
pac	packing
per	perforation
pha	pneumatic hammers
pla	planes
pli	pliers and tongs
pol	polishing
pri	printing
pul	pullers
rea	reamers
riv	riveting
rwy	railways
saf	safety appliances
san	sanders
saw	saws and sawing
scd	screwdrivers
sci	scissors
scr	screws and screwthreads
sep	separators
sha	sharpening
she	shearing
soc	social activities
spa	spanners
spo	sports
sto	stone working
sur	surveying
tex	textiles
too	tools
tra	transport
vic	vices
wat	watchmaking
wel	welding
woo	woodworking

GB	English, British usage
US	English, American usage
v	verb
f	Français
e	Español
i	Italiano
n	Nederlands
d	Deutsch

Basic table

A

1 **aboutsledge; sledge; sledge hammer**
[ham]
f marteau à devant
e macho; martillo de dos manos
i martello a due mani; martello da battere
n voorhamer
d Vorschlaghammer

2 **abrasive band** *(cloth)*; **abrasive belt** *(cloth)*
[gri/san]
f bande émerisée *(toile)*; ruban émerisé *(toile)*
e cinta esmeril *(lienzo)*; correa esmeril *(lienzo)*
i cinghia abrasiva *(tela)*; nastro abrasivo *(tela)*
n bandschuurlinnen
d Bandschleifleinen

3 **abrasive band** *(paper)*; **abrasive belt** *(paper)*
[gri/san]
f bande émerisée *(papier)*; ruban émerisé *(papier)*
e cinta esmeril *(papel)*; correa esmeril *(papel)*
i cinghia abrasiva *(carta)*; nastro abrasivo *(carta)*
n bandschuurpapier
d Bandschleifpapier

* **abrasive belt** *(cloth)* → 2

4 **abrasive cloth**
[gri/san]
f toile abrasive
e lienzo abrasivo
i tela abrasiva
n schuurlinnen
d Schleifleinen

5 **abrasive disk; sanding disk**
[gri/san]
f disque abrasif
e disco abrasivo
i disco abrasivo
n schuurschijf
d Schleifscheibe

6 **abrasive paper**
[gri/san]
f papier abrasif
e papel abrasivo
i carta abrasiva
n schuurpapier
d Schleifpapier

7 **abrasive ring; sanding ring**
[cpt/san]
f anneau abrasif
e anillo abrasivo
i anello abrasivo
n schuurring
d Schleifring

8 **abrasive roll; sanding roll**
[gri/san]
f rouleau d'atelier
e rollo de cinta abrasiva
i rotolo a nastro abrasivo
n spaarrol
d Schmirgelschnur; Sparrolle

9 **abrasives**
[gri/pol/san]
f abrasifs
e abrasivos
i abrasivi
n slijpmiddelen
d Schleifmittel

10 **abrasive wheel dresser; grinder scouring serrated wheels**
[gri/san]
f décrasse-meule; molettes emmanchées pour décrassage de meules
e alisador de muela; rectificador de muelas
i equilibratore di mole; lisciatore di mola
n opwerker; slijpsteenscherper
d Schleifscheibenabrichter

11 **abutment cheek; collar; peg shoulder**
[car/joi/woo]
f épaulement de tenon
e hombro de muñón
i spalla di tenone
n borst
d Zapfenschulter

12 **acetylene blowpipe**
[bur]
f chalumeau à acétylène
e soplete de acetileno
i cannello ad acetilene
n acetyleenbrander
d Azetylenbrenner

13 **acid pump**
[che]
f pompe à acides
e bomba de ácidos
i pompa ad acidi

n zuurpomp
d Säurepumpe

14 adhesive trowel; mastic trowel; serrated-edge trowel
[mas]
f truelle à côté(s) denté(s)
e llana dentada
i cazzuola dentata
n getande plakspaan
d gezahnte Kelle

15 adjustable bar
[agr/gen]
f barre de réglage
e varilla de ajuste
i barra regolabile
n regelstaaf
d Stelleisen

16 adjustable bow
[agr]
f étrier; joug; travée
e puente; yugo
i arco di regolazione
n regeljuk
d Brücke; Joch; Stellbogen; Stellbügel

17 adjustable ga(u)ge
[mea]
f calibre ajustable
e calibre ajustable
i calibro aggiustabile
n bijstelkaliber; nastelbaar kaliber
d nachstellbare Lehre

18 adjustable level
[mea/sur]
f niveau ajustable
e nivel ajustable
i livella aggiustabile
n bijstelbare waterpas
d nachstellbare Wasserwaage

19 adjustable pulling claws
[cpt/pul]
f griffes extensibles
e garras prolongables
i griffe aggiustabili
n verstelbare trekhaken
d verstellbare Abzughaken

20 adjustable reamer
[car/dri/rea/woo]
f alésoir ajustable; broche ajustable
e alisadora ajustable; brocha ajustable
i alesatore regolabile; broccia regolabile
n verstelbare broots; verstelbare ruimdoorn
d nachstellbare Räumnadel

21 adjustable screw
[scr]
f vis réglable
e tornillo ajustable
i vite regolabile
n verstelbare schroef
d Verstellschraube

22 adjustable spanner; screw wrench
[spa]
f clé à crémaillère
e llave ajustable
i chiave a rullino
n schroefsleutel
d Rollgabelschlüssel

23 adjustable wire stripping pliers
[pli]
f pince à dénuder à vis
e alicate(s) pelacables ajustable(s)
i pinza spellafili regolabile a vite
n instelbare afstriptang
d einstellbare Abisolierzange

24 adjusting lever
[tra]
f levier compensateur
e palanca compensadora; palanca igualadora
i leva di registro
n compensatiehefboom
d Ausgleichhebel

25 adjusting nut
[scr]
f écrou de réglage
e tuerca de ajuste; tuerca de enclavamiento
i dado di registrazione
n bijstelmoer
d Nachstellmutter; Stellmutter

26 adjusting pin
[gen]
f cheville de montage
e pasador de alentas
i asta di regolazione
n centerpen; stelpen
d Stellstift

27 adjusting pliers
[pli]
f pince d'ajustage
e alicate(s) de ajuste
i pinza di aggiustaggio
n justeertang
d Justierzange

28 adjusting roll
[gen]

f rouleau d'ajustage
e rodillo de ajuste
i rullo di aggiustaggio
n afstelrol
d Justierrolle

29 adjusting screw
[gen/scr]
f vis calante; vis d'arrêt; vis de réglage
e tornillo de ajuste
i vite di registrazione
n stelschroef
d Klemmschraube; Spannschraube; Stellschraube

30 adze
[agr/car/coo/wool]
f herminette
e azuela de mano
i ascia; mazzapicchio
n dissel; gootdissel
d Dexel; Hohldexel

31 aerograph; air brush; paint spray; spray gun
[dec]
f pistolet de vernissage; pulvérisateur pour peinture
e pistolete aerográfico
i aerografo; atomizzatore di colori; spruzzatore per vernice
n verfspuit
d Farbspritze; Farbspritzpistole

32 agitator
[che]
f agitateur; brasseur
e agitadora
i agitatore
n roertoestel
d Rührgerät

33 agricultural tools; farmer's tools
[agr]
f outils aratoires
e aperos; herramientas de agricultores
i arnesi agricoli
n akkergereedschap; akkertuig
d landwirtschaftliche Werkzeuge

34 aiguille
[dri]
f aiguille
e aguja
i ago
n boornaald; steenboortje
d Bohrnadel

* **air brush → 31**

35 air chuck
[cpt/lat]
f mandrin de serrage pneumatique
e mandril neumático; plato neumático
i mandrino pneumatico
n pneumatische spanklauw
d Druckluftspanndorn; Pressluftspannfutter

36 air compressor
[hyd]
f compresseur d'air
e compresora de aire
i compressore d'aria
n luchtcompressor
d Luftverdichter

37 air drill
[dri]
f perforatrice à air comprimé
e taladro neumático
i trapano pneumatico
n luchtdrukboor; pneumatische boor
d Pressluftbohrer

38 air hammer; compressed air hammer; pneumatic hammer; pneumatic power hammer
[ham]
f marteau pneumatique
e martillo neumático
i martello pneumatico
n pneumatische hamer
d Drucklufthammer

39 alidade; sight rule
[meal]
f alidade
e alidada
i alidada
n alhidade; vizierliniaal
d Alhidade

40 all-geared head drilling machine
[dri]
f perceuse rapide à colonne et activation directe
e taladro rápido con mando por caja de velocidades y monopolea
i trapano rapido a colonna e ad attivazione diretta
n direct aangedreven kolomboormachine
d Säulenbohrmaschine mit Direktantrieb

41 alligator spanner; alligator wrench; cylinder spanner; cylinder wrench
[spa]
f clé crocodile
e llave de mordazas paralelas
i chiave a ganasce per tubi

n pijpsleutel
d Rohrschlüssel

42 all-purpose circular saw blade
[cpt/saw]
f lame multimatériaux de scie circulaire
e hoja de sierra circular universal
i lama di sega circolare universale
n universeel cirkelzaagblad
d Allzweckkreissägeblatt

43 all-purpose saw
[saw]
f scie universelle
e sierra universal
i sega universale
n universeelzaag
d Allzwecksäge

44 all-purpose scissors; universal scissors
[sci]
f ciseaux universels
e tijeras universales
i cesoie universali
n allesknipper; combischaar
d Allzweckschere; Universalschere

45 all-steel wood chisel
[chi]
f ciseau à bois tout en acier
e cincel para madera todo de acero
i scalpello da legno tutto acciaio
n stalen houtbeitel
d stählerner Holzbeitel

46 alternate hardpoint sharpening
[cpt/saw]
f affûtage croisé
e afilado cruzado
i affilatura incrociata
n gekruiste slijpmethode
d gekreuztes Schleifverfahren

47 alternating toothing
[saw]
f denture alternée
e dentadura alternante
i dentatura alternante
n afwisselend naar weerszijden gezette vertanding
d wechselseitig geschränkte Zahnung

48 altitude level
[mea]
f niveau d'altitude
e nivel de altitud
i livella di altitudine
n hoogtewaterpas
d Höhenwasserwaage

49 aluminium oxide paper; corundum paper
[gri/san]
f papier à corindon; papier à oxyde d'aluminium
e papel de corindón; papel revestido de alúmina sintética
i carta a corindone; carta ad alumina sintetica
n korundpapier; synthetisch aluminiumoxyde-papier
d Aluminiumoxydschleifpapier; Korundschleifpapier

50 amateur tools; DIY-tools; do-it-yourself tools; hobby tools
[gen]
f outils d'amateur; outils de bricoleur
e herramientas para el aficionado; herramientas para el bricolaje
i utensileria di hobbisti
n doe-het-zelver-gereedschap
d Bastlerwerkzeuge; Heimarbeitwerkzeuge; Hobbywerkzeuge

51 American toothing
[saw]
f denture américaine
e dentadura americana
i dentatura americana
n Amerikaanse vertanding
d amerikanische Zahnung

52 angle block
[car/joi/woo]
f renfort
e taco de ángulo
i blocco di rinforzo
n versterkingsblokje
d Verstärkungsblock

53 angle board
[car/joi/woo]
f support cornier
e soporte angular
i supporto angolare
n hoeksteun
d Winkelstütze

54 angle brace; angle tie
[car/joi/woo]
f contrefiche
e codillo de ángulo
i rinforzo angolare
n hoekverband
d Schwenkbug; Winkelband

55 angle cutting plane
[pla]
f rabot pour angles

e cepillo para ángulos
i pialletto per angoli
n hoekschaaf
d Winkelhobel

56 angled painting brush; radiator brush
[dec]
f pinceau à radiateurs
e pincel para radiadores
i pennello per radiatori
n bokspoot; bokkepoot; radiatorkwast
d Heizkörperpinsel

57 angle drill
[cpt/dri]
f foret cornier
e broca de ángulo
i trapano ad angolo
n hoekboor
d Winkelbohrer

58 angle drilling attachment
[dri]
f adaptation renvoi d'angle de perceuse
e aditamento de broca angular
i dispositivo a trapano angolare
n hoekbooraanzetstuk
d Winkelbohrvorsatz

59 angle float; triangular trowel
[mas]
f truelle triangulaire
e trulla triangular
i cazzuola triangolare
n driehoekige troffel
d Dreieckkelle

60 angle grinder
[gri]
f meuleuse d'angle
e esmerilador angular
i smerigliatrice angolare
n haakse slijpmachine
d Winkelschleifer

61 angle iron shears
[met/she]
f cisailles à cornières
e tijeras de hierro ángulo
i cesoie per il taglio di ferri ad angolo
n hoekijzerschaar
d Winkeleisenschere

62 angle lever
[civ]
f genouillière; levier coudé
e palanca acodada
i leva ad angolo
n knie; kniehefboom
d Kniehebel

63 angle meter
[mea]
f mesureur d'angles
e medidor de ángulos
i misuratore di angoli
n hoekmeter
d Winkelmesser

*** angle tie → 54**

64 angle trowel; plasterer's twitcher
[mas]
f pelle de plâtrier
e paletada de estucador
i pialletto di stuccatore
n stukadoorsschepje
d Stukkateurkasten

65 angular cutter
[fre]
f fraise angulaire
e fresa angular
i fresa ad angolo
n hoekfrees
d Winkelfräser

66 angular drilling machine
[dri]
f perceuse angulaire
e taladro angular
i trapano angolare
n haakse boormachine
d Winkelbohrmaschine

67 angular ga(u)ge
[mea]
f jauge angulaire
e calibre angular
i calibro angolare
n hoekkaliber
d Winkelkaliber; Winkellehre

68 angular reamer
[dri/fre/rea]
f alésoir à équerre; alésoir angulaire
e escariador angular
i alesatore angolare
n haakse handruimer
d Winkelaufreiber; Winkelreibahle

69 annular auger; annular bit
[dri]
f foret annulaire
e taladro anular
i trapano anulare

n ringboor
d Kreisbohrer; Ringbohrer

70 **annular borer**
[dri/min]
f tête foreuse annulaire
e barrena hueca
i perforatore a nucleo scorrevole
n kernboor
d Kernbohrgerät

71 **anvil**
[for/met]
f enclume
e yunque
i incudine
n aambeeld; aanbeeld
d Amboss

72 **anvil bed**
[for/ham]
f chabotte de marteau-pilon
e chabota de martinete a vapor
i incudine inferiore
n stoomhamerblok
d Chabotte; Unteramboss

73 **anvil block; anvil cushion**
[cpt/for]
f billot d'enclume
e cepo de yunque
i ceppo d'incudine
n aambeeldsblok; aambeeldsstoel; aambeeldsstok; aanbeeldsblok; aanbeeldsstoel; aanbeeldsstok
d Ambossklotz; Ambossstock; Ambossuntersatz

74 **anvil body**
[cpt/for]
f pied d'enclume
e pie de yunque
i piede d'incudine
n aambeeldsvoet; aanbeeldsvoet
d Ambossfuss

75 **anvil chisel; anvil cutter; hatchet stake**
[cpt/for]
f tranchet d'enclume (à queue)
e tajadera de yunque
i tagliolo d'incudine
n schrooi
d Ambossabschröter

76 **anvil cone**
[cpt/for]
f cône d'enclume
e cono de yunque
i cuneo da incudine
n aambeeldhoorn; aanbeeldhoorn
d Ambosshörnchen

* **anvil cushion → 73**

* **anvil cutter → 75**

77 **anvil face; anvil plate**
[cpt/for]
f face d'enclume; table d'enclume
e mesa de yunque; plano de yunque; tabla de yunque
i piano di lavoro d'incudine
n baan van een aanbeeld; bovenvlak van een aanbeeld
d Ambossbahn

78 **anvil horn; beak iron; beck iron; bickern; bick iron**
[cpt/for]
f bigorne; corne d'enclume
e bigornia; cuerno de yunque
i corno di incudine
n aambeeldhoorn; aanbeeldhoorn
d Ambosshorn

* **anvil plate → 77**

79 **apron; wedge**
[pla]
f coin; languette
e cuña
i cuneo
n keg; spie
d Keil

80 **arbor**
[cpt/gen]
f arbre; mandrin
e árbol; mandril
i albero; mandrino
n spil
d Spindel; Welle

81 **arbor for cutter; cutter arbor; miller arbor**
[fre]
f mandrin portefraises
e eje portafresas
i mandrino portafrese
n freesspil
d Frässpindel

82 **archimedean drill**
[dri]
f drille
e berbiquí de carrete
i trapano ad archetto; trapano di Archimede

n drilboor
d Drillbohrer

83 arch punch; wad punch
[lea]
f emporte-pièce à arche
e sacabocado(s) de arco
i fustella ad arco
n holpijp met beugel
d Henkellocheisen

84 arm file; coarse file; rough file
[far]
f lime à bras; lime grosse; lime rude
e lima carretela gruesa; limatón
i lima grossa
n armvijl; grove vijl
d Armfeile; Bestossfeile; Grobfeile

85 arrow
[mea/sur]
f flèche d'arpenteur
e aguja de cadenero; bastín marcador
i biffa; freccia indicatrice
n landmetershaak; merkstok
d Markierstab

86 articulated bit drilling machine
[dri]
f perceuse à broches articulées
e taladro de husillos articulados
i trapano a mandrini articulati
n boormachine met gelede spillen
d Gelenkspindelbohrmaschine

87 artificial manure distributor
[agr/hor]
f distributeur d'engrais chimiques
e repartidor de abonos artificiales
i spandiconcime
n kunstmeststrooier
d Kunstdüngerstreuer

88 asbestos gloves
[gen]
f gants en amiante
e guantes de amianto
i guanti di amianto
n asbesthandschoenen
d Asbesthandschuhe

89 asbestos washer
[gen]
f rondelle en amiante
e arandela de amianto
i rondella di amianto
n asbestsluitring
d Asbestunterlegscheibe

90 asparago knife; asparagus cutter
[agr]
f coupe-asperges
e cortaespárragos; cuchillo esparraguero
i coltello per asparagi
n aspergemes
d Spargelmesser

91 asphalt rake
[civ/tra]
f râteau d'asphalte
e rastrillo de asfalto
i rastrello da asfaltare
n asfalthark
d Asphaltharke; Asphaltrechen

92 astragal plane
[pla]
f rabot d'astragale
e cepillo de astrágalo
i pialla di astragalo
n astragaalschaaf
d Astragalhobel

93 astragal tools
[pla]
f outils pour astragales
e herramientas para astrágalo
i utensileria per astragalo
n astragaalgereedschap
d Astragalwerkzeuge

94 atomizer; nebulizer
[agr/foy/hor]
f pulvérisateur; vaporisateur
e nebulizador; pulverizador
i atomizzatore; irroratrice; nebulizzatore
n nevelverstuiver; vernevelspuit
d Zerstäuber

95 attachment screw
[scr]
f vis de fixation
e tornillo de sujeción
i vite di fissaggio
n bevestigingsschroef
d Befestigungsschraube

96 auctioneer's hammer; gavel
[ham]
f marteau de commissaire-priseur
e martillo de rematador
i martello di banditore
n afslaghamer
d Auktionatorhammer

97 auger; wimble
[car/dri/woo]
f tarière

e barrena
i trivella
n agger; avegaar; effer
d Stangenschlangenbohrer

98 auger bit; twist bit
[car/dri/woo]
f mèche à hélice unique
e barrena salomónica; broca helicoidal
i mecchia a tortiglione
n slangenboor; slingerboor
d Schlangenbohrer

99 auger lead screw
[car/dri/woo]
f pointe de tarière
e punta de barrena
i punta di trivella
n avegaarschroefpunt
d Stangenbohrerspitze

100 autolock chuck; automatic locking chuck; self-tightening chuck
[cpt/dri]
f mandrin à serrage automatique; mandrin autoserrant
e mandril de bloqueo automático; mandril de enclavamiento automático
i mandrino a bloccaggio automatico; portapunta a chiusura automatica
n zelfspannende boorhouder
d Schnellspannfutter; Selbstspannfutter

101 autolock cutter
[fre]
f fraise à blocage automatique
e fresa de bloqueo automático
i fresa a bloccaggio automatico
n zelfspannende frees
d Selbstspannfräser; selbstspannender Fräser

102 automatic drilling machine
[dri]
f perceuse automatique
e taladro automático
i trapano automatico
n boorautomaat
d Bohrautomat

103 automatic guiding chain
[civ/tra]
f chaîne d'attelage; chaîne-guide
e cadena para la guía automática
i catena per la guida automatica
n leiketting
d Führungskette; Selbsthaltekette

104 automatic level
[mea/sur]
f niveau automatique
e nivel automático
i livella automatica
n automatische waterpas
d selbsttätige Wasserwaage

* **automatic locking chuck → 100**

105 automatic marker punch
[cpt/mea]
f pointeau automatique à repérer
e punzón automático para marcar
i punzone automatico per marcare
n automatische markeerpons
d selbsttätiger Markierkörner

106 automatic nailing machine
[bui/car]
f appareil automatique de clouage
e aparato automático para clavar
i apparecchio automatico per chiodare
n spijkerautomaat
d selbsttätiger Nagelungsapparat

107 automatic screwdriver; spiral ratchet screwdriver
[scd]
f tournevis automatique
e destornillador automático
i cacciavite automatico
n pompschroevedraaier
d Drillschrauber

108 automatic wire stripping pliers
[pli]
f pince automatique à dénuder
e alicate(s) pelacables automático(s)
i pinza spellafili automatica
n automatische afstriptang
d automatische Abisolierzange

109 averuncator; long-handled pruner; tree pruner
[agr/foy/hor]
f échenilloir; élagueur
e podadera de varilla; tijera de ramas
i svettatoio
n boomschaar
d Astschere; Baumschere

110 awl
[per]
f alène
e lezna
i punteruolo
n els; priem
d Ahle; Pfrieme

111 **axe**
[axe]
f hache
e hacha
i ascia
n aaks; aks; akst; bijl
d Axt; Beil

112 **axe blade**
[axe]
f fer de la hache
e hoja de la hacha
i lama dell'ascia
n bijlblad
d Axtblatt

113 **axe eye**
[axe]
f œil de la hache
e ojo de la hacha
i occhio dell'ascia
n bijloog
d Axtöhr

114 **axe handle; axe helve**
[axe]
f manche de la hache
e astil de la hacha; mango de la hacha
i manico dell'ascia
n bijlsteel
d Axtstiel; Beilhelm; Beilstiel

115 **axe head**
[axe]
f talon de la hache
e cabeza de la hacha
i testa dell'ascia
n bijlhuis
d Axthaupt; Axthaus

* **axe helve → 114**

116 **axle**
[cpt/gen]
f axe
e eje
i asse
n as
d Achse; Welle

117 **axle stand**
[axe]
f support d'axe
e soporte de eje
i supporto di asse
n assteun
d Achsenstütze; Wellenbock

B

118 back
[cpt/saw]
f dos
e dorso
i dorso
n rug
d Rücken

119 back
[cpt/saw]
f dosseret
e costilla
i costola
n zaagkap; zaaglade
d Rückenschiene

120 back; beck
[agr/houl]
f cuve
e cuba
i botte; mastello
n kuip
d Bottich

121 back bench-screw
[car/cpt]
f boîte de rappel; presse de derrière; vis de derrière
e pinza trasera
i morsa posteriore di banco
n achterklauw; achtertang
d Hinterzange

122 backing pad
[cpt/iwa]
f plateau souple
e anillo de inserción
i anello d'inserimento
n steunflens; steunschijf
d Einlegering

123 backsaw; miter(re) saw; tenon saw
[saw]
f scie à dos
e sierra de dorso
i sega a dorso
n kapzaag
d Rückensäge

124 back square; try square
[car/mea]
f équerre épaulée
e escuadra con espaldón
i squadra a cappello
n blokhaak; blokwinkelhaak; schrijfhaak
d Anschlagwinkel; Tischlerwinkel

125 badger
[mas]
f déchaussoir; racloir
e rastrillo
i lisciatoio; utensile per scolmare
n afstrijker
d Abstreifer

126 badger hair
[dec]
f pinceau en poil de blaireau
e pincel en pelos de tejón
i pennello di peli di tasso
n dassehaarpenseel
d Dachshaarpinsel

127 badger hair softener
[dec]
f blaireau
e suavizador de pelo de tejón
i spazzola di tasso
n daskwast
d Dachsvertreiber

128 badger plane
[pla]
f guillaume incliné
e guillame de esquinar; guillame inclinado
i sponderuola inclinata
n boordschaaf
d schräger Simshobel

129 bailer
[nav]
f sasse
e achicador; vertedor
i gottazza; sessola
n hoosvat
d Ösfass; Schöpfeimer

130 baker's shovel; battledore; peel
[nut]
f pelle à enfourner
e pala de enhornar; pala del pan
i infornapane; pala per infornare
n bakkersschop; bakkersschotel; schieter; schietplank
d Brotschieber

131 bale
[pac/tra]
f balle
e bala; fardo
i balla
n baal
d Ballen

132 bale breaker
[pac/tra]

f déballeur
e abrebalas; abridor de balas
i apriballe
n baalbreker
d Ballenöffner

133 bale elevator; hay elevator
[pac/tra]
f élévateur à balles
e elevador-estibador de balas
i elevatore di balle
n balentransporteur
d Ballenstapler-Förderer

134 bale hook
[pac/tra]
f crochet à balle
e garfio de bala
i gancio di balla
n baalhaak
d Ballenhaken

135 bale loader; hay loader
[agr/tra]
f chargeur de balles
e cargadora de pacas
i caricatore di balle
n balenlader
d Ballenlader

136 baler
[agr/tra]
f presse à emballer
e enfarderladura; prensa para pacas de paja
i macchina per imballare
n balenpers
d Ballenpresse

137 ballast hammer
[ham]
f maille casse-pierres; masse à carrer; massette pour cantonniers
e maceta; porrillo de cantero
i martello schiacciapetri
n steenslaghamer
d Steinschlägel; Steinschlaghammer

138 ballast shovel
[civ]
f pelle à fratras
e pala de lastre
i pala di zavorra
n ballaster; ballastschop; bats; schepbats; steekbats
d Schrotschaufel

139 ball bearing
[cpt/gen]
f butée à billes; palier à billes; roulement à billes
e cojinete de bolillas; rulemán
i cuscinetto a sfere
n kogellager
d Kugellager

140 ball bearing puller
[pull]
f extracteur de coussinets; extracteur de roulement à billes
e extractor de cojinetes
i estrattore per cuscinetti
n kogellagertrekker
d Kugellagerabzieher

141 ball bearing retaining ring puller; race puller
[pull]
f extracteur de la bague de roulement à billes
e extractor del anillo de retención del rodamiento de bolas
i estrattore dell'anello di cuscinetto a sfere; estrattore della ralla
n kogellagerloopringtrekker
d Kugellagerlaufringabzieher

142 ball joint puller
[pull]
f extracteur de joint à rotule
e extractor de junta de rótula
i estrattore di giunto sferico
n kogelgewricht
d Kugelgelenkabzieher

143 ball pein hammer
[ham]
f marteau à boule anglais
e martillo de bola
i martello a penna sferica; martello a presella tonda
n bolbankhamer
d Ballhammer

144 ball race
[cpt/joi]
f bague de roulement à billes; cage à billes
e anillo del rodamiento de bolas
i ralla
n kogellagerloopring
d Kugellagerlaufring

145 ball stake
[cpt/for]
f bigorne sphérique; enclumette sphérique
e bigornia esférica
i bicornia sferica; incudinetto sferico

n bolstaak
d ballförmige Faust

146 band clutch
[joi]
f embrayage à ruban
e embrague de cinta
i frizione a nastro
n wrijvingskoppeling met band
d Bandkupplung; Reibbandkupplung

147 band facer
[gri/pol]
f ruban émerisé
e cinta abrasiva; correa lijadora
i cinghia per smerigliare; nastro per smerigliare
n schuurband
d Schleifband

148 band grinder; band sander; belt grinder; belt sander
[gri/san]
f meuleuse à bande; meuleuse à ruban
e esmerilador de cinta; esmerilador de correa
i smerigliatrice a cinghia; smerigliatrice a nastro
n bandschuurmachine
d Bandschleifmaschine

149 bandsaw
[saw]
f scie à ruban
e sierra de cinta
i sega a nastro
n bandzaag; lintzaag
d Bandsäge

150 bandsaw attachment
[saw]
f adaptation scie à ruban
e aditamento para sierra de cinta
i adattamento per sega a nastro
n lintzaagvoorzetstuk
d Bandsägevorsatz

151 bandsaw blade
[cpt/saw]
f lame de scie à ruban sans fin
e hoja de sierra de cinta sin fin
i lama di sega a nastro senza fine; nastro senza fine
n zaaglint
d endloses Sägeblatt

152 bandsaw machine
[saw]
f machine à scie de ruban
e máquina de sierra de cinta
i macchina a sega a nastro
n bandzaagmachine
d Bandsägemaschine

153 band steel scissors
[iwa/she]
f ciseaux pour feuillard
e tijeras para hierro en fleje
i cesoie per reggetta
n bandstaalschaar
d Bandstahlschere

154 band steel tensioning apparatus
[iwa]
f appareil de serrage de feuillard
e aparato tensor de hierro en fleje
i apparecchio tensitore di reggetta
n bandstaalspanapparaat
d Bandstahlspanngerät

155 bank
[agr/hor]
f faux de rive
e guadaña de orilla
i falce da rive
n hakoord; walzeis
d Grabenrandsense

156 bar cramp
[car/joi]
f serre-joint
e sargento
i sergente
n lijmknecht; sergeant; serre-joint
d Schraubzwinge; Spannknecht

157 bark brush; tree brush
[foy]
f brosse-émoussoir
e cepillo descortezador
i spazzola per scortecciare
n boomborstel; boomschorsborstel
d Baumbürste; Rindenbürste

158 bark ga(u)ge
[foy]
f calibre d'épaisseur d'écorce
e calibre de espesor de corteza
i calibro per spessori di corteccia
n schorsdiktemeter
d Rindedickenmesser

159 barking drum
[for]
f tambour à décortication
e tambor giratorio descortezador; tambor para descortezar
i tamburo per scortecciare

n ontschorsingstrommel
d Entrindungstrommel

* **barking iron** → 161

160 **barking spade; bark spud**
[foy]
f décortiqueur; pelle à écarcer
e escoplo para descortezar
i pala per scortecciare
n boomkat; boommes; haalmes
d Rindenschälmesser; Schäleisen; Schnitzmesser; Ziehmesser; Zugmesser

161 **bark scaler; barking iron; peeling iron**
[foy]
f tranchant coupe-écorce
e descortezador de hierro
i arnese per scortecciare
n schilmes
d Loheisen; Lohlöffel; Rindenschäler

162 **bark scraper; tree scraper**
[foy]
f émoussoir
e raspador de árboles; raspador de corteza
i raschietto per alberi
n boomkrabber
d Baumkratzer; Rindenkratzer

* **bark spud** → 160

163 **barrel; vat**
[che/gen]
f baril; fût; tonneau
e barril; tonel
i barile; botte
n fust; ton; vat
d Fass; Tonne

164 **barrel-shaped hammer; barrel-shaped mallet**
[car/ham]
f maillet en bois en forme de baril; marteau en bois en forme de barillet
e martillo de madera en forma de barril; mazo de madera en forma de tonel
i martello di legno in forma di barilotto; mazzuola di legno in forma di barile
n houten hamer in tonmodel; tonhamer
d tonnenförmiger Holzhammer

165 **barrette file**
[far]
f barrette; lime barrette
e lima triangular achatada
i lima triangolare appiattata
n afgeplatte driekante vijl
d Barrett-Feile

166 **barrow; wheelbarrow**
[agr/foy/hor]
f brouette
e carretilla
i carriola
n kruiwagen
d Schiebkarre(n); Schubkarre(n)

167 **basin wrench**
[hyd/pli]
f clé pour lavabo
e llave para lavabo
i chiave per lavandino
n kraanmoersleutel; wastafelmoersleutel
d Waschbeckenschlüssel

168 **bastard cut**
[far]
f taille bâtarde
e picadura de bastarda
i taglio bastardo
n bastordhak
d Bastardhieb; Hieb 1

169 **batch box; ga(u)ge box**
[mas/mea]
f auge de mesure; boîte jaugée
e cajón medidor
i cassa di misura
n maatbak
d Messkasten

170 **bath wrench**
[hyd/pli]
f clé pour baignoire
e llave para bañera
i chiave per tinozza
n badkuipsleutel
d Badewanneschlüssel

171 **battery filler**
[ele/tra]
f pipette
e sifón
i sifone di aspirazione
n accuvloeistofvuller
d Saugheber

172 **battery hydrometer**
[ele/mea/tra]
f pèse-acide
e acidímetro
i acidimetro
n accuzuurmeter
d Säuremesser

173 **battery terminals puller**
[ele/pul/tra]
f extracteur de cosses de batterie

e extractor de bornes de batería
i estrattore di morsetti di batteria
n accupoolklemtrekker
d Batterieklemmenabzieher

174 battery tester
[ele/mea/tra]
f essayeur de batteries
e probador de baterías; verificador de baterías
i collaudatore di batterie; forcella di controllo degli elementi
n accuceltester
d Akkumulatorenprüfer; Akkuzellenprüfer; Batteriezellenprüfer

* **battledore → 130**

175 bayonet saw; heavy-duty saw; reciprocating saw; sabre saw
[saw]
f scie à mouvement alternant
e sierra alternativa
i sega a movimento alterno
n elektrische schrobzaag
d elektrische Stichsäge

176 beading bit; beading cutter
[car/fre/woo]
f fraise 1/4 sphérique
e fresa 1/4 esférica
i fresa 1/4 sferica
n kwartrondfrees
d Viertelstabfräser

177 bead plane; beading plane
[car/pla]
f rabot à boudins
e cepillo de junquillos
i sponderuola per listelli
n kraalschaaf
d Rundstabhobel; Simshobel

178 bead saw
[saw]
f petite scie à dosseret
e pequeño serrucho de costilla
i piccola sega a dorso
n kleine rugzaag
d kleine Rückensäge

179 bead-tool
[for]
f bigorne pour façonner les bourrelets
e bigornia para formar bordes
i scalpello a doccia
n kleine speerhaak
d Sickenstock

180 beak
[cpt/pla]
f pointe de patte
e apoyo
i punta del barletto
n houvastkrul
d Hafterspitze

181 beak iron; beck iron; bick iron; bickern
[for]
f corne d'enclume
e cuerno de yunque
i corno di incudine
n aambeeldhoorn; aanbeeldhoorn
d Ambosshorn

182 beak iron; beck iron; bickern; bick iron
[cpt/for]
f bigorne
e bigorneta
i bicornia
n speerhaak
d Sperrhaken

* **beak iron → 78**

183 beam clamp
[bui/cla/woo]
f serre-poutre
e abrazadera para viga; pinza-viga
i pinza-trave
n balkenklem
d Balkenklemme

184 beam compasses; beam trammel; trammel
[mea]
f compas à verge
e compás de varas
i compasso a verga
n stokpasser
d Stangenzirkel

185 bearing
[cpt/gen]
f coussinet; palier
e cojinete; rodamiento
i cuscinetto
n lager
d Lager

186 bearing puller
[pull]
f extracteur de palier
e extractor de cojinetes; extractor de rodamiento
i estrattore per cuscinetti
n lagertrekker
d Lagerabzieher

* **beck → 120**

* **beck iron → 78, 181, 182**

187 beetle; maul
[ham]
f maillet; mailloche
e martinete; mazo
i mazza; mazzuolo
n beukhamer; moker
d Bleuel; Schlegel

188 beet lifter
[agr]
f fourche arracheuse à betteraves
e arrancador de remolachas
i cavatuberi
n beetwortellifter
d Rübenhebegabel

189 belaying cleat; belaying pin
[iwa/nav]
f cabillot; chevillot; taquet
e cabilla; cornamusa
i caviglia
n belegnagel; korvijnagel
d Belegnagel

190 bell center(re) punch
[per]
f perçoir automatique
e granete automático; punzón autoconcentrador
i punzone autocentrante
n automatische centerpons; automatische keurnagel; automatische kornagel
d selbsttätiger Ankörner

191 bell-crank lever
[tra]
f levier à renvoi; levier coudé
e palanca acodada; palanca en escuadra
i leva ad angolo retto
n kniehefboom
d Kniehebel; Winkelhebel

192 bell-mouthed bore
[dri]
f trou de forage émarginé
e agujero de perforación usado
i foro di trivellazione svasaro
n uitgesleten boorgat
d ausgezacktes Bohrloch

193 bellows
[for/gen]
f soufflet
e fuelle
i soffietto
n blaasbalg
d Blasebalg

194 belt
[gen/tra]
f courroie
e correa
i cinghia
n riem
d Riemen

195 belt fastener
[tra]
f agrafe de courroie; attache-courroie
e abrazadera de correas; grapa para correas
i graffa per cinghie
n riemkram; riemverbinder
d Riemenklammer; Riemenverbinder

196 belt fork; belt striker
[tra]
f fourchette de courroie
e horquilla de correas; horquilla de desembrague
i spostacinghia
n riemvork
d Riemengabel

* **belt grinder → 148**

197 belt pulley
[tra]
f poulie à courroie
e polea para correa
i puleggia a cinghia
n riemschijf
d Riemenscheibe; Transmissionsscheibe

108 belt punch; hollow punch; saddler's hollow punch
[gen/lea]
f emporte-pièce rond
e punzón sacabocados; sacabocado(s) para correa
i fustella per guarnizioni
n holpijp
d Locheisen

* **belt sander → 148**

* **belt striker → 196**

199 bench; workbench; working table
[gen]
f banc; établi
e banco de trabajo; mesa de trabajo
i banco da lavoro

n werkbank
d Arbeitstisch; Werkbank; Werktisch

200 **bench anvil; bench block**
[for]
f enclumeau; enclumette; tasseau d'établi
e yunque de banco
i incudine di banco
n bankaambeeld
d Bankamboss

201 **bench brush**
[gas]
f brosse
e escobilla de mesa
i pennello per pulire il tavolo
n bankborstel
d Brettbürste

202 **bench chisel**
[chi]
f ciseau d'établi
e escoplo de banco
i scalpello di banco
n bankbeitel
d Bankbeitel

203 **bench clamp; bench holdfast**
[car/cpt/gen]
f valet d'établi
e prisionero de banco
i barletto
n werkbankklem
d Werktischklemme

204 **bench drilling machine**
[dri]
f perceuse à table; perceuse d'établi
e taladro de mesa; taladro para banco
i trapano da banco
n tafelboormachine
d Tischbohrmaschine; Werkbank-bohrmaschine

205 **bench grinder; bench sander**
[gri/san]
f meuleuse d'établi; touret d'établi
e esmerilador de mano; esmerilador de mesa
i smerigliatrice da banco
n handslijpmachine; werkbankslijpmachine
d Handschleifmaschine; Tischschleif-maschine

206 **bench hammer; engineer's hammer**
[ham]
f marteau d'établi
e martillo de banco
i martello da banco
n Engelse bankhamer
d Bankhammer

207 **bench-hand's hammer**
[ham]
f marteau d'ajusteur
e martillo de ajustador
i martello per meccanico
n bankwerkershamer
d Schlosserhammer

208 **bench-hand's tools**
[cpt/gen]
f outillage d'ajusteur
e herramientas de ajustador
i utensileria di falegname; utensileria di lavoratore al banco
n bankwerkersgereedschap
d Schlosserwerkzeuge

* **bench holdfast → 203**

209 **bench hook**
[car/cpt/pla]
f presse d'établi
e tope del banco
i cane del banco
n bankklauw; banktang; stoothaak
d Bankenzange

210 **bench hook; bench stop**
[cpt/gen]
f crochet; griffe; mentonnet; valet d'établi
e barrilete; gancho del banco; tope del banco
i barletto; cane del banco; gancio del banco
n bankhaak; klemhaak
d Bankeisen; Bankhaken

211 **bench pan; goldbeater's pan**
[gas]
f bassin d'établi
e caldero de batidor de oro
i cassetta raccogli-limatura
n schrootlade
d Feilungsblech

212 **bench plane**
[car/pla/woo]
f rabot plat d'établi
e cepillo de banco
i pialla da banco; pialletto da banco
n bankschaaf
d Bankhobel

213 **bench rule**
[mea]
f perche de mesure en bois; règle divisée en bois

e regla dividida en madera
i pertica da misurare di legno
n houten meetlat
d hölzerne Messlatte

* **bench sander → 205**

214 bench shears; stock shears
[she]
f cisailles à banc à levier
e tijeras de banco
i troncatrice a mano con doppia trasmissione a leva
n bankschaar
d Stockschere mit Hebelübersetzung

215 bench skin; goldbeater's skin
[gas]
f baudruche
e piel de batidor de oro
i buccio di battiloro
n goudslagershuidje; goudslagersvel
d Goldschlägerhäutchen

216 bench stone; whetstone
[sha]
f pierre à affiler; pierre à aiguiser
e piedra de afilar
i pietra da affilare
n vlakke slijpsteen; wetsteen
d Honstein; Wetzstein

* **bench stop → 210**

217 bench vice ***GB*****; bench vise** ***US***
[cla/vic]
f étau d'établi
e tornillo de banco
i morsa parallela
n bankschroef
d Bankschraubstock; Parallelschraubstock

218 bending apparatus
[ben]
f appareil à cintrer; appareil à plier
e aparato doblador; aparato plegador
i apparecchio per curvare; apparecchio per piegare
n buigapparaat
d Biegeapparat

219 bending ga(u)ge
[ben/mea]
f gabarit de courbure; gabarit de flexion
e galga dobladora
i maschera da piegare
n buigmal; buigschoen
d Biegelehre; Biegeschablone

220 bending iron; bending pin
[ben/met]
f fer à couder
e barra para doblar; herramienta para abocardar
i mordiglione
n buigijzer
d Biegeeisen

221 bending machine
[ben]
f plieuse
e máquina dobladora; máquina plegadora
i curvatrice; piegatrice
n buigmachine
d Biegemaschine

* **bending pin → 220**

222 bending springs
[ben/iwa]
f ressorts de flexion
e muelles de flexión
i molle a flessione
n buigveren
d Biegefedern

223 bent beak iron
[for]
f tas coudé
e tas de espiga curvado
i tassello ricurvo
n bolle tas; gebogen tas
d gebogener Tasso

224 bent chisel; dog-leg chisel
[chi]
f butte-avant; ciseau courbé
e cincel acodado; cincel curvo
i scalpello ricurvo
n gebogen beitel
d gekrümmter Beitel

225 bent gouge
[chi]
f gouge courbée
e gubia acodada; gubia curva
i sgorbia ricurva
n gebogen guts
d gekrümmter Hohlbeitel

226 bent lateral stake
[cpt/for]
f fer à rabattre (rond); tranche ronde de ferblanterie
e bigornia lateral acodada
i bicornia laterale curva
n gebogen randstaak
d Umschlageisen mit runder Bahn

227 **bent long snipe nose pliers**
[pli]
f pince à becs longs affilés et coudés
e alicate(s) de bocas largas puntiagudas y curvadas
i pinza a becchi lunghi appuntati e curvi
n tang met lange spits uitlopende gebogen bekken
d Langbeckzange mit abgebogenen spitz zulaufenden Backen

228 **bent round beck iron**
[cpt/for]
f tasseau rond bombé
e tas redondo
i tassello curvato
n gebogen tas
d gebogener Tasso

229 **bent snipe nose pliers**
[pli]
f pince à becs affilés et coudés
e alicate(s) de bocas puntiagudas y curvadas
i pinza a becchi appuntati e curvi
n tang met spits uitlopende gebogen bekken
d Zange mit abgebogenen spitz zulaufenden Backen

230 **betty; chairmaker's saw**
[hou/saw]
f scie à arc de chaisier
e sierra de arco de sillero
i sega ad arco di seggiolaio
n stoelenmakerszaag
d Bügelsäge eines Stuhlmachers

231 **bevel edge plane**
[car/pla]
f rabot à plates-bandes
e cepillo de listón moldurado
i pialla per listelli
n bandschaaf; bossingschaaf
d Randleistenhobel

232 **bevel gear**
[tra]
f engrenage conique
e engranaje cónico
i ingranaggio conico
n kegeltandwieloverbrenging
d Kegelradübersetzung

233 **bevel(l)ed edge**
[chi/gen]
f bord chanfreiné
e arista biselada
i canto smussato
n afgeschuinde kant; vouw
d abgeschrägte Kante

234 **bevel(l)ed-edge chisel**
[chi]
f ciseau à bords biseautés
e cincel de bordes biselados
i scalpello a bordi smussati
n beitel met afgeschuinde kanten
d Beitel mit abgeschrägten Kanten

235 **bevel(l)ed edges**
[car/chi/gen]
f chanfreins
e bordes biselados; chaflanes
i fili smussati
n schuine zijkanten
d Schrägkanten

236 **bevel(l)ed point flat file**
[far]
f lime plate galbée
e lima plana de punta biselada
i lima piana a punta smussata
n platte vijl met afgeschuinde punt
d Flachfeile mit Spitzenverjüngung

237 **bevel-sharpened teeth**
[cpt/saw]
f denture à affûtage biseauté; denture à affûtage croisé
e dentadura de afilado biselado; dentadura de afilado cruzado
i dentatura ad affilatura a bisellatura; dentatura ad affilatura incrociata
n vertanding met schuine afwerking
d Zahnung mit Kreuzschliff; Zahnung mit Schrägschliff

238 **bias cutting**
[saw]
f coupe en biais
e cortado al bies; cortado al sesgo
i taglio diagonale
n schuine snede
d Schrägschnitt

* **bickern → 78, 181, 182**

* **bick iron → 78, 181, 182**

239 **billhook; branchbill; lopping billhook**
[for/hor]
f couperet; émondoir; serpe
e bodollo; podón
i coltella; mezzaluna; roncola per sfrondare
n bijlmes; groot snoeimes; kapmes
d Gertel; Hackmesser; Haumesser; Heppe

240 binding nail
[fas/iwa]
f clou à tête de plomb
e clavo con cabeza de plomo
i chiodo a testa di piombo
n bindnagel; loodkopnagel
d Bindenagel; Dachdeckernagel

241 bit
[cpt/hou/saf]
f panneton
e paletón
i ingegno
n baard
d Bart

242 bit; drill bit; grooved bit
[dri]
f foret; mèche
e barrena; broca
i mecchia
n boor
d Bohrer

243 bit adapter
[dri]
f adaptateur de mèche
e adaptador de barrena; adaptador de broca
i adattatore di mecchia
n booropnamestuk
d Bohreraufnahmegerät

244 bit clamp
[cpt/hou]
f serre de panneton
e pinza de paletón
i pinza d'ingegno
n baardklem
d Bartklemme

245 bit ga(u)ge; bit stop
[dri]
f arrêt de mèche
e tope de barrena; tope de broca
i arresto per mecchia
n booraanslag
d Bohreranschlag

246 bit point
[dri]
f pointe de mèche
e punta de barrena; punta de broca
i punta di mecchia
n boorpunt; boorspits
d Bohrerspitze

247 bit shank
[cpt/dri]
f queue de mèche; tige de mèche
e caña de barrena; caña de broca
i stelo di mecchia
n boorkolf
d Bohrerschaft

248 bit stock; brace; drill brace
[dri]
f vilebrequin
e berbiquí
i girabecchina; menarola
n boorbeugel; booromslag
d Bohrkurbel; Bohrleier; Bohrwinde; Brustleier; Faustleier

* **bit stop → 245**

249 blacksmith
[for]
f forgeron
e forjador
i fabbro
n smid
d Schmied

250 blacksmith's mate
[for]
f aide-forgeron; frappeur
e ayudante de forjador; batidor
i battimazza; garzone di fabbro
n smidsknecht; voorslager
d Schmiedegeselle; Zuschläger

251 blacksmith's tongs
[for/pli]
f tricoise
e alicate(s) de herrero
i tenaglia di fabbro ferraio
n hoeftang
d Hufbeschlagzange

252 blade bit; drag bit; fishtail bit
[dri/min]
f trépan à queue de poisson
e barrena de cola de pescado
i perforatore a coda di pesce
n visstaartbeitel
d Fischschwanzbohrer

253 blade of a file
[cpt/far]
f plaque de lime
e hoja de lima
i paletta di lima
n vijlblad
d Feilenkörper

254 blind mortise; stopped mortise
[car/joi/woo]

f mortaise aveugle; mortaise sans issue
e mortaja ciega
i mortasa cieca
n blind gat
d blindes Zapfenloch

255 blind mortise and tenon joint; blind tenon and mortise joint
[car/joi/wool]
f assemblage aveugle à mortaise et tenon; assemblage sans issue à mortaise et tenon
e ensambladura ciega de mortaja y muñón
i incastro cieco a mortasa e tenone
n blinde pen- en gatverbinding
d blinde Verzapfung

256 blind rivet
[riv]
f rivet faux
e remache ciego
i chiodo cieco; ribattino cieco
n blindklinknagel
d Blindniete

257 blind riveting pliers; pop riveter; riveter
[pli/riv]
f pince à rivets; tenaille(s) de rivetage faux
e tenazas para remaches ciegos; tenazas para roblones
i tenaglie per chiodatura cieca
n blindklinktang; popklimtang; poptang
d Blindnietzange; Nietzange

258 blind riveting tools
[riv]
f outils de rivetage faux
e herramientas para remaches ciegos
i arnesi per chiodatura cieca
n blindklinkgereedschap
d Blindnietwerkzeuge

*** blind tenon and mortise joint → 255**

259 block; pulley block
[tra]
f moufle; palan
e motón
i bozzello
n blok; takelblok
d Block; Takelblock

260 block and pulley set; block and tackle
[tra]
f palan
e garrucha; polipasto; polispasto
i paranco
n takel
d Flaschenzug; Takel

261 block ga(u)ge
[mea]
f cale-étalon; calibre à limites parallèles
e bloquito calibrador; calibre prismático
i calibro a blocco parallelo
n eindmaat
d Endmass; Parallelendmass; Vorrichtungsformlehre

262 block plane
[pla]
f rabot d'établi; rabot de finissage
e cepillo handyman
i pialla handyman; pialletto per spianare di testa
n blokschaaf; zoetschaaf
d Putzhobel; Schlichthobel

263 blowing iron; blowing tube; blowpipe
[gla]
f canne de verrier; fêle de souffleur; felle de souffleur
e caña de vidriero; tubo de soplador
i canna di soffiatore; canna di vetraio
n blaaspijp; glasblazerspijp
d Glasbläserpfeife

264 blowlamp *GB*; blowtorch *US*
[wel]
f lampe à souder
e lámpara de soldar
i lampada per saldare
n soldeerlamp
d Lötlampe

*** blowpipe → 263**

265 blunt chisel
[chi]
f ciseau émoussé; ciseau obtus
e cincel embotado; cincel romo
i scalpello ottuso; scalpello smussato
n botte beitel
d stumpfer Beitel

266 blunt file
[far]
f lime obtuse
e lima de dientes romos; lima embotada
i lima ottusa
n botte vijl
d abgestumpfte Feile

267 blunt start
[scr]
f pointe non-filetée d'une vis
e punta no-roscada de un tornillo; supresión del hilo de rosca al principio de ésta

i punta non filettata di una vite
n gladde schroefpunt
d Schraubenspitze ohne Gewinde

268 board drop hammer; gravity drop hammer
[ham]
f marteau-pilon à planche
e martillo de tabla
i maglio a tavola
n plankhamer; valhamer met plank
d Brettfallhammer

269 boarding axe
[nav]
f hache d'abordage
e hacha de abordaje
i azzo d'abbordaggio
n enterbijl
d Enterbeil

270 boaster; boasting chisel
[chi]
f ébauchoir
e cincel ancho
i scalpello largo
n brede beitel
d Breitmeissel

271 bodkin
[pri]
f pointe
e punta
i pinza
n letterzetterspriem
d Punkturspitze

272 bodkin
[lea/tex]
f aiguille à passer; passe-cordon; passe-lacet
e aguja de jaretas; punzón de satre
i ago de tappezziere
n rijgpen
d Durchzieh-Schnürnadel

273 boilermaker's hammer
[ham]
f marteau de chaudronnier
e martillo de calderero
i martello di calderaio
n ketelmakershamer
d Kesselschmiedehammer

274 boiler reamer
[dri/fre/rea]
f alésoir de chaudronnier
e escariador de calderas
i alesatore di calderaio
n ketelmakersruimer
d Kesselräumer

275 boiler scaling hammer; scaling hammer
[ham]
f marteau à piquer les chaudières
e martillo para picar calderas
i martello di nettatore di caldaio
n ketelbikhamer; ketelsteenbikhamer
d Kesselsteinhammer

276 bolt
[gen/iwa]
f boulon
e perno
i bullone
n bout
d Bolzen

277 bolt
[hou/iwa/saf]
f verrou
e cerrojo; colanilla; pestillo
i chiavarda; perno passante
n grendel
d Riegel

278 bolt chisel; bolting iron; drawer lock chisel; lock bolt chisel
[chi]
f ciseau à grignoter
e cortafrío de lengua de carpa
i scalpello per fare cavi per la stanghetta
n hoekbeitel
d Aushauer; Lochbeitel

279 bolt cutters
[iwa/pli/she]
f coupe-boulons
e cortapernos
i tagliabulloni
n boutenschaar
d Bolzenschneider

280 bolt drift; bolt driving punch
[gen/iwa]
f chasse-boulons
e botador de pernos
i cacciabulloni
n boutdrevel; boutdrijfijzer
d Bolzendupeisen; Bolzentreiber

281 bolted connection
[car/joi/woo]
f assemblage par boulons; boulonnage
e unión empernada
i giuntura a mezzo di bulloni
n boutverbinding
d Bolzenverbindung

282 bolt extractor
[gen/iwa/pul]
f arracheur de boulons; chasse-boulons
e sacapernos
i cacciabulloni; estrattore di bulloni
n boutentrekker; boutuitdraaier
d Bolzenauszieher

* **bolting iron → 278**

283 bolt spanner; bolt wrench
[iwa/spa]
f clé de boulons
e llave para pernos
i chiave per bulloni
n boutsleutel
d Bolzenschlüssel

284 bone saw
[nut/saw]
f scie à os
e serrucho de carnicero
i sega da ossa
n slagerszaag
d Knochensäge

285 border plane
[car/pla]
f rabot à traiter les bords
e cepillo de cantear
i sponderuola a bastone
n randschaaf
d Kantenhobel

286 border shears
[hor]
f coupe-bordures
e tijeras para cortar bordes de césped
i forbici tosaerba per bordura
n randjesknipper
d Rasenrandschere

287 bore
[cpt/gen]
f alésage
e agujero cilíndrico; diámetro interior; pared interior
i alessaggio; diametro interno
n asgat; inwendige diameter
d Höhlung; innerer Zylinderdurchmesser

288 bore chips; bore dust
[dri]
f alésures; copeaux de forage; poussière de foret
e virutas de taladrado
i trucioli di trapanatura
n boorsel; boorspanen
d Bohrspäne

289 bore rod; boring bar
[dri]
f arbre d'alésage; arbre de foret; tige de foret
e barra de taladrar
i barra di alesaggio
n boorstang
d Bohrstange

290 boring machine
[dri]
f perceuse
e taladradora
i trapano
n boormachine
d Bohrmaschine

291 boring mill
[dri]
f perceuse à mèche fixe
e taladradora de broca fija
i trapano a mecchia fissa
n boormachine met vaststaande boor
d Bohrmaschine mit feststehender Bohrspitze

292 boring tool
[dri]
f outil à percer
e herramientas taladradoras
i arnesi da trapano; utensile da trapano
n boorgereedschap
d Bohrwerkzeug

293 bossing mallets
[ham/hyd]
f mallets du plombier
e martillos del plomero; mazos del fontanero
i martelli del lattoniere
n loodgietershamers; loodkloppers
d Bleihämmer; Klempnerschmiedehämmer

294 bottle brush
[cle/hou/nut]
f écouvillon; goupillon; rince-bouteille
e escobillón para limpiar botellas
i spazzola per bottiglie
n flesseborstel
d Flaschenbürste

295 bottle jack
[tra]
f vérin en forme de bouteille
e gato de tornillo; gato en forma de botella
i martinetto; verricello a vite
n flesvijzel
d Schraubenflaschenzug

296 bottom die
[for]
f matrice inférieure
e matriz inferior
i matrice inferiore
n ondermatrijs
d unterer Schmiedesattel

297 bottoming tap; finishing tap; plug tap; third tap
[dri/scr]
f taraud finisseur
e macho acabador; macho de roscar final
i maschio finitore
n nasnijder
d Fertigschneider

298 bottom knife
[coo]
f couteau de fond
e cuchillo de fondo
i coltello da fondo
n bodemmes
d Bodenmesser

299 bottom reamer
[dri/fre/rea]
f alésoir de fond
e escariador de fondo
i alesatore di fondo
n bodemruimer
d Grundreibahle

300 bottom rim plane
[coo]
f rabot à chanfreiner
e acepillador de cantear
i pialletto per bisellatura
n bodemrandschaaf
d Kantenhobel

301 bottom swage
[cpt/for]
f étampe dessous à queue
e estampa inferior
i stampo inferiore
n onderstempel
d Rundgesenk unten; Untergesenk

302 bottom wood
[coo]
f bois de fond
e madera de fondo
i legno di fondo
n bodemhout
d Bodenholz

303 bow; brace; clip; strap
[iwa]
f bride; étrier
e estribo
i archetto; staffa
n beugel
d Bügel

304 bow
[cpt/saw]
f arc
e arco
i arco
n beugel; boog
d Bogen; Bügel

305 bow cal(l)ipers; cal(l)iper compasses; egg cal(l)ipers; outer compasses; outside cal(l)ipers
[mea]
f compas d'épaisseur; compas extérieur
e compás de espesores; compás de gruesos
i compasso di spessore; compasso per esterno
n buitenpasser; krompasser
d Aussenzirkel; Greifzirkel; Tastzirkel

306 bow drill; drill borer
[dri]
f drille
e berbiquí helicoidal; parahúso
i trapano ad archetto
n drilboor; slakkeboor; spilboor
d Drillbohrer

307 bow saw
[saw]
f scie à arc; scie à bûches
e sierra de arco
i sega ad arco
n beugelzaag; boogzaag
d Bogensäge; Bügelsäge

308 box angle plate
[cpt/gen]
f support cubique cornier
e apoyo cúbico angular
i supporto cubico angolare
n opspankubus
d Aufspannwürfel

309 box of punches
[cpt/per]
f boîte à poinçons
e caja de punzones
i scatola dei punzoni
n ponsbus
d Punzenbüchse

310 box screwdriver
[scd]

f tournevis à douille
e destornillador de tubo
i cacciavite a collare
n dopschroevedraaier
d Steckschraubendreher

311 box spanner
[spa]
f clé à canon; clé à douille; clé à tube
e llave de tubo; llave de vaso
i chiave a collare; chiave a tubo
n dopsleutel
d Rohrschlüssel; Steckschlüssel

312 box spanner component parts
[cpt/spa]
f accessoires de clés à canon; accessoires de clés à douille
e piezas accesorias de llaves de tubo
i accessori per chiavi a collare; accessori per chiavi a tubo
n dopsleutelhulpstukken
d Steckschlüssel-Zubehörteile

313 brace; wimble
[dri/sto]
f vilebrequin
e berbiquí
i menarola
n booromslag
d Bohrwinde

*** brace → 248, 303**

314 brace jaws
[dri]
f mâchoires de vilebrequin
e mordazas de berbiquí
i mandrini di menarola
n boorhouderbekken
d Bohrwindenbacken

315 brace screwdriver
[scd]
f tournevis à vilebrequin
e destornillador de berbiquí
i cacciavite a menarola
n booromslagenschroevedraaier
d Bohrwindenschraubendreher

316 brad
[iwa]
f clou à tête perdu
e clavo sin cabeza
i chiodo senza testa
n koploze spijker
d Nagel ohne Kopf

317 bradawl
[per]
f alène plate
e lezna plana
i punteruolo a lama piatta
n platte Engelse priem
d flache Ahle

318 bradawl; sprig bit
[cpt/per]
f alène plate; alène pour les clous; poinçon
e punzón plano
i punteruolo per piantare chiodi
n nagelpriem; platte els
d Bindahle; flache Ahle; Nagelort

319 brake adjusting spanner
[spa/tra]
f clé pour ajustage du frein
e llave para ajustar el freno
i chiave per aggiustaggio del freno
n remstelsleutel
d Bremsstellschlüssel

320 braking iron; kneading iron
[nut]
f brisoir
e herramienta de amasar
i gramola per impastare
n braakijzer
d Kneteisen

*** branchbill → 239**

321 branch cutting axe
[agr/foy/hor]
f cognée à branches
e hacha de desrame
i ascia per tagliare i rami
n takkenbijl
d Ästeaxt

322 branding iron
[agr]
f fer à marquer
e ferrete
i marchio
n brandijzer; schroeiijzer
d Brandeisen; Brenneisen

323 brazing hearth
[cpt/wel]
f foyer à soudure forte
e fogón para soldadura fuerte; hogar para soldadura fuerte
i focolare a lega per brasatura; forno a lega per brasatura
n hardsoldeeroven
d Hartlötofen

324 brazing table
[cpt/wel]
f table à soudure forte
e mesa para soldadura fuerte
i tavola per brasatura
n hardsoldeertafel
d Hartlöttisch

325 brazing torch
[wel]
f fer à soudure forte
e soldador para soldadura fuerte
i arnese per brasatura
n hardsoldeerbout
d Hartlötkolben

326 breakdown crane *GB*; **wrecking crane** *US*
[tra]
f grue de dépannage
e grúa de avaría
i carro gru di soccorso
n ophaalkraanwagen
d Abschleppkran

327 breakdown lamp
[cpt/tra]
f lampe de dépannage
e lámpara de avería
i lampada di guasto
n pechlamp
d Pannelampe

328 breakdown repair set
[cpt/tra]
f coffret de dépannage
e cofre de eliminación de avería
i borsa attrezzi per guasto
n pechkoffer
d Pannenbeseitigungskoffer

329 break-in *v* **a new file**
[far]
f roder une lime
e rodar una lima
i rodare una lima
n een vijl inrijden
d eine Feile einlaufen

330 breast; mo(u)ldboard
[mas]
f oreille; versoir
e orejera; vertedera
i orecchio; versoio
n strijkbord
d Streichbrett

331 breast drill
[dri]
f chignole; porte-foret à conscience
e berbiquí de pecho
i trapano a petto
n borstboor
d Brustbohrgerät; Brustleier

332 brick axe; bricklayer's hammer; tiling hammer
[ham/mas]
f marteau de maçon; martelet
e alcostana; piqueta; zapapico
i martello da muratore
n kaphamer; metselaarshamer
d Maurerhammer

333 brick bolster; brick chisel
[chi/sto]
f ciseau à brique; ciseau de tailleur
e cortaladrillos
i scalpello da mattoni
n steenhouwersbeitel
d Steinspaltmeissel

334 brick hammer; bricklayer's hammer
[civ/ham]
f marteau à brique
e martillo de enladrillar
i martello di stradino
n kaphamer
d Pflasterhammer; Steinsetzerhammer

335 brick jointer; jointer; slicker; striking tool
[mas]
f fer à joint de brique; truelle à joints de briques
e paletín
i ferro per giunti di mattoni
n voegspijker
d Fugenkelle

336 bricklayer's bolster; mason's bolster
[mas]
f hachard
e cortafrío de albañil
i scalpello a frio del muratore
n ceseel
d breitschneidiger Steinmeissel

337 bricklayer's chisel
[chi/mas]
f ciseau de maçon
e cortafrío de albañil
i scalpello di muratore
n voegbeitel
d Mauermeissel

* **bricklayer's hammer → 332, 334**

338 bricklayer's knife
[mas]
f couteau de maçon
e cuchillo de albañil
i coltello di muratore
n voegmes
d Fugenmesser

339 brick trowel; buttering trowel; laying trowel
[mas]
f truelle de maçon
e paleta de albañil
i cazzuola di muratore
n metselaarstroffel
d Maurerkelle

340 brier-tooth; gullet; gullet tooth
[cpt/saw]
f dent de loup
e espacio entre dientes
i dente di lupo
n ronde spatie; wolfstand
d Einschweifung der Sägezähne; Wolfszahn

341 brier-tooth saw; gullet saw
[saw]
f scie à dents de loup
e sierra con dientes de lobo
i sega a denti di lupo
n wolfstandzaag
d Wolfszahnsäge

342 broach
[dri/fre]
f broche
e brocha
i broccia
n broots; freesschaaf; trekfrees
d Räumahle; Räumnadel; Reibahle

343 broaching chisel
[chi/met/real]
f ciseau alésoir
e cincel brochador
i scalpello alesatore
n brootsbeitel; kotterbeitel; ruimbeitel
d Räumbeitel

344 broaching file
[far]
f lime à brochage
e lima de brocha; lima de brochado
i lima per brocciare; lima per spinare
n brootsvijl
d Räumfeile

345 broad; turner's chisel
[chi/lat]
f biseau
e escoplo de torneros
i scalpello per tornitori
n draaibeitel; slechtbeitel
d Drehmeissel

346 broad axe
[axe/car]
f hache de charpentier
e hacha de carpintero
i scure a taglio largo
n timmermansbijl
d Breitbeil

347 broad axe
[foy]
f hache de bûcheron
e hacha de armas; hacha de leñador
i ascia da tagliaboschi
n houthakkersbijl
d Holzfällersbeil

348 broadcast sower
[agr]
f semeuse
e sembradora
i seminatrice a spargimento
n zaadstrooier
d Sämaschine

349 broad flat brush
[dec]
f queue de morue épaisse
e brocha plana ancha
i pennello piatto largo
n slechtkwast
d Flachpinsel

350 broom
[cle/gen/hou]
f balai à manche
e escoba de mano
i scopa
n handbezem
d Handbesen

351 broomstick
[cle/gen/hou]
f manche à balai
e palo de la escoba
i manico della scopa
n bezemsteel
d Besenstiel

352 brush
[dec]
f brosse; pineau
e pincel
i pennello

n kwast
d Pinsel

353 brush with weight
[cle/hou]
f hérisson avec boulot
e peso esférico para golpear
i spazzola con peso
n schoorsteenvegersragebol
d Sonne mit Kugelschlagapparat

354 bruzze; grooving tool
[chi/lat]
f burin de tourneur en V
e acalanador en V; escoplo de tornero en V
i scalpello di tornitore in V
n V-vormige steekbeitel
d Abstechstahl; Plattenstahl

355 bucket; pail
[dre/gen/hou]
f godet; seau
e balde; cangilón; cubo de agua
i secchia
n emmer
d Eimer

356 bucking saw; one-man cross-cut saw
[foy/saw]
f scie à tronçonner
e sierra tronzadora
i segaccio a mano con telaio
n eenmanstrekzaag
d Ablängsäge; Trennsäge

357 buckle; metal strap
[fas/iwa]
f bride métallique; étrier métallique
e estribo metálico
i nastro metallico; staffa
n metaalbeugel
d Metallbügel

358 budding knife
[agr/hor]
f greffoir
e navaja de injertar
i coltello per innesto
n oculeermes
d Okuliermesser

359 buff stick
[pol/wat]
f astic; buffle; cabron; polissoir
e costa
i bisegolo; bussetto
n likbeen; poleerstok; polijststok
d Glättholz; Polierstift

360 buff wheel
[pol]
f meule à polir; polissoir; polisseuse
e disco polidor
i disco di panno per polire
n met leer overtrokken polijstschijf
d Polierscheibe

361 buhl saw
[saw]
f scie à grande bûche
e sierra de gran arco
i sega a grande arco
n spanzaag met hoog freem
d Spannsäge mit hohem Bügel

362 building tools
[bui/gen/too]
f outils de construction
e herramientas de construcción
i utensileria di costruzione
n bouwgereedschap
d Bauwerkzeuge

363 bull nose rebate plane; edge plane
[pla]
f guillaume à fer en bout; guillaume d'ébéniste
e cepillo corte de cuello ajustable y frente separable
i sponderuola a smusso
n neusschaaf
d Möbeltischler-Simshobel

364 bung ***GB*****; faucet** ***US***
[coo]
f bondon
e bitoque; tapón
i zaffo
n bom; spon
d Spund

365 bung hole ***GB*****; faucet hole** ***US***
[coo]
f bonde
e boca de barril
i cocchiume
n bomgat; spongat
d Spundloch

366 bung hole borer ***GB*****; faucet hole borer** ***US***
[coo]
f bondonnière; jarbière; tarière
e abonador; taponadora
i succhiello di bottaio
n bomboor; kuipersboor; sponboor
d Spundbohrer; Stuhlbohrer

367 buoy tongs
[nav]
f tenaille(s) de bouée
e tenazas de boyas
i tenaglie di boe
n boeitang
d Bojenzange

368 burin; graving tool
[art/chi]
f burin rond; traceret
e buril de grabador
i bulino rotondo
n burein; burijn; graveernaald
d Grabstichel; Gravierstichel; Radiernadel

369 burner
[bur]
f chalumeau
e mechero
i becco; cannello
n brander
d Brenner

370 burner reamer
[bur/dri/fre/rea]
f alésoir pour brûleur
e escariador para quemador
i alesatore per bruciatore
n branderruimer
d Brennerreibahle

371 burring reamer
[fre/met/rea]
f alésoir à dégrossir; alésoir d'ébarbage; alésoir ébaucheur
e escariador desbarbador; escariador de desbaste
i alesatore sbavattore
n afbraamruimer; voorruimer
d Schruppreibahle; Vorreibahle

372 burs; rotary burs; rotary milled files and rasps
[far/fre/pol]
f fraises rotatives
e fresas giratorias
i frese rotanti
n rotatiefrezen; roterende raspen en vijlen; stiftfrezen
d rotierende Fräser; Rotorfräser

373 bush knife; machete; matchet
[foy]
f grand cutelas; machette; sabre d'abattis
e machete
i coltellaccio
n groot kapmes; machete
d grosses Hackmesser; Waldmesser

374 butcher's axe
[nut]
f hachette
e hachuela de carnicero
i mannaia
n hakbijl; slagersbijl
d Hackbeil

375 butcher's hook
[nut]
f croc à viande
e gancho de carnicero; gancho para colgar carne
i uncino di sospensione
n vleeshaak
d Aufhängebügel

376 butcher's saw
[nut/saw]
f scie de boucher
e sierra de carnicero
i sega da macellaio
n vleeszaag
d Fleischersäge; Knochensäge

377 butter drill
[nut]
f flûte; perce-beurre
e taladro de mantequilla
i trapano di burro
n boterboor
d Butterbohr

378 butterfly nut; flynut; wing nut
[iwa]
f écrou à oreilles
e tuerca mariposa
i dado a farfalla; dado alettato; galletto
n bladmoer; vleugelmoer
d Flügelmutter

* **buttering trowel → 339**

379 butt ga(u)ge; feeler ga(u)ge
[mea]
f espion
e calibrador de separaciones; galga de rendijas
i calibro a lama
n scharniermeter
d Füllerlehre; Spion

380 butt joint
[car/joi/woo]
f assemblage à plat
e junta a tope; junta plana
i giunto di testa
n koud-tegen-elkaar
d Endverbindung; Stossverbindung

381 **button-headed screw; half-round screw**
[iwa/scr]
f vis à tête en goutte de suif; vis à tête lentiforme
e tornillo de cabeza lenticular
i vite con testa a calotta
n schroef met bolverzonken kop
d Linsenkopfschraube

382 **buzz saw** *US*; **circular saw** *GB*
[saw]
f scie circulaire
e sierra circular
i sega a disco; sega circolare
n cirkelzaag
d Kreissäge

383 **buzz saw axis** *US*; **circular saw axis** *GB*
[cpt/saw]
f arbre de scie circulaire
e árbol de sierra circular
i albero di sega circolare
n cirkelzaagas
d Kreissägenachse

384 **buzz saw bank** *US*; **circular saw bank** *GB*
[saw]
f banc à scie circulaire
e banco de sierra circular
i banco a sega circolare
n cirkelbank; cirkelzaagbank
d Kreissägebank

385 **buzz saw blade** *US*; **circular saw blade** *GB*
[cpt/saw]
f lame de scie circulaire
e hoja de sierra de disco
i lama di sega circolare
n zaagschijf
d rundes Sägeblatt

386 **buzz saw file** *US*; **circular saw file** *GB*
[cpt/saw]
f lime pour scie circulaire
e lima de sierra circular
i lima per sega circolare
n cirkelzaagvijl
d Kreissägenfeile

C

387 cabinet file
[far]
f lime d'ébéniste
e lima de ebanista
i lima per ebanista
n kabinetvijl
d Kabinettfeile

388 cabinet maker
[car/woo]
f ébéniste
e ebanista
i ebanista
n schrijnwerker
d Kunsttischler

389 cabinet rasp
[far]
f râpe d'ébéniste
e escofina de ebanista
i raspa per ebanista
n kabinetrasp
d Kabinettraspel

390 cable drill; churn drill; percussion drill; pneumatic drill
[dri/min/pha]
f marteau d'abattage; marteau-piqueur pneumatique
e martillo picador neumático; perforadora neumática
i martello pneumatico; trivella a percussione
n luchtdrukhamer; pneumatische hamer
d Abbauhammer

391 cable reeling unit
[gen/too]
f prolongateur-enrouleur
e tambor de cable
i avvolgicavo a mano
n kabelhaspel
d Kabeltrommel

392 cable saw; wire saw
[saw]
f scie à pierre; scie de carrière; scie passepartout; sciotte
e sierra al hilo
i sega a filo
n draadzaag; kabelzaag; marmerzaag; steenzaag
d Drahtsäge; Steinsäge

393 cable tongs
[ele/pli]
f pinces pour câbles
e alicate(s) para cables
i pinza per cavi
n kabeltang
d Kabelzange

394 calibration board; calibration table; calibration test bench
[mea]
f banc d'essai d'étalonnage
e banco de prueba de calibración
i banco di prova di taratura; quadro per taratura
n ijkproefbank
d Eichprüfstand

395 cal(l)iper compasses; outer compasses
[mea]
f compas d'épaisseur; compas extérieur
e compás de exteriores
i compasso di spessore; compasso per esterno
n buitenpasser
d Aussenzirkel; Greifzirkel; Tastzirkel

* **cal(l)iper compasses → 305**

396 cal(l)iper ga(u)ge
[mea]
f jauge réglable
e calibre plano
i calibro piatto
n speermaat
d Tasterlehre

397 cambrel; gambrel
[nut]
f jambier
e curvatón; esparrancador; estirador
i uncino divaricatore
n slachtersboom
d Krummholz; Spriegel

398 Canterbury hoe; potato hoe
[agr/hor]
f croc à pommes de terre
e azada para las patatas
i zappa scavapatate
n aardappelschoffel
d Kartoffelhacke

399 cant hook; peavy
[foy]
f grappin; renard; tourne-bille
e gafa de palanca; palanca de gancho
i asta a gancio; gancio per travi
n balkhaak; duivelsklauw; kanthaak
d Kanthaken; Stammwender; Wendehaken

400 cape chisel; chasing chisel; cross-cut chisel
[chi/met/sto]
f ciseau à rainurer
e cincel ranurador; cortafrío de noja estrecha para ranurar
i scalpello stretto per scanalature
n kantbeitel; ritsbeitel
d Nutenmeissel; Rilleisen

401 cap iron lever; cap lever
[cpt/pla]
f levier de bloc d'arrêt
e leva de bloqueo de la cuña
i leva di bloccaggio del ferro
n dekbeitelhefboom
d Exzenterhebel der Messerklappe

402 cap screw
[iwa/scr]
f vis à tête
e tornillo de cabeza
i vite con testa
n kopschroef
d Kopfschraube; Schraubbolzen mit Rechteckkopf

403 capstan wheel nut spanner; spider spanner; spider wrench
[spa]
f clé à pipe multiple; clé en croix
e llave de cruz; llave de vaso múltiple
i chiave a bussola multipla; chiave a croce
n kruissleutel; meervoudige dopsleutel
d Kreuzschlüssel; Mehrfachsteckschlüssel

404 carbide cutting edge; carbide tip
[cut/gen/met/too]
f plaquette de coupe en métal dur
e plaquita de metal extraduro
i tagliente al carburo
n hardmetaalsnijkant
d Hartmetallbestückung

405 carbide tipped cutter
[fre]
f fraise au carbure
e fresa con cuchillas de carburo de tungsteno
i fresa al carburo di tungsteno
n frees met hardmetaallaag
d hartmetallbestückter Fräser

406 carbide tipped saw blade
[cpt/saw]
f lame de scie à plaquettes de coupe en métal dur
e hoja de sierra con plaquita de metal extraduro
i lama di sega a punte di carburo
n zaagblad met hardmetaalsnijvlakjes
d Sägeblatt mit Hartmetallbestückung

407 carborundum wheel; silicon carbon wheel
[cpt/gri/san]
f meule au carbure de silicium
e muela de carborundo
i mola a carborundum
n carborundumslijpschijf; siliciumcarbideslijpschijf
d Karborundumschleifscheibe; Siliziumkarbidschleifscheibe

408 carcass saw
[saw]
f scie grosse à dos
e sierra gruesa de dorso
i sega grossa a dorso
n grove rugzaag
d grobe Rückensäge

409 card nipper
[hou/pli]
f perforateur à main
e perforador de mano
i perforatore a mano
n handperforator
d Handlochgerät

410 card puncher
[per]
f perforateur mécanique
e perforador mecánico
i perforatore meccanico
n mechanische perforator
d maschinelles Lochgerät

411 car jack
[tra]
f cric pour voitures; lève-auto
e gato para automóviles
i cricco di automobile; martinetto di automobile
n autokrik; autolichter
d Wagenheber

412 carpenter
[car/gen]
f charpentier; menuisier
e carpintero
i carpentiere; falegname
n timmerman
d Zimmermann

413 carpenter's hammer
[car/ham]
f marteau de charpentier; marteau de

menuisier
e martillo de carpintero
i martello di carpentiere; martello di falegname
n timmermanshamer
d Zimmermannshammer

414 **carpenter's level**
[car/mea]
f niveau de charpentier; niveau de menuisier
e nivel de carpintero
i livella di carpentiere; livella di falegname
n timmermanswaterpas
d Schreinerwasserwaage; Tischlerwasserwaage; Zimmermannswasserwaage

415 **carpenter's mallet; joiner's mallet; square mallet**
[car/ham]
f maillet carré de charpentier; maillet carré de menuisier
e mazo cuadrado de carpintero
i mazzuolo quadrangolare di carpentiere; mazzuolo quadrangolare di falegname
n klopper; houten timmermanshamer
d Zimmermannsklüpfel

416 **carpenter's pencil**
[car/wool]
f crayon de charpentier; crayon de menuisier
e lápiz de carpintero
i matita di carpentiere; matita di falegname
n timmermanspotlood
d Zimmermannsbleistift

417 **carpenter's steel square; flat square; framing square; rafter square; roofing square**
[car/mea/wool]
f équerre de charpentier; équerre de menuisier
e escuadra de carpintero
i squadra di acciaio di carpentiere; squadra di acciaio di falegname
n timmermanshaak; winkelhaak
d Zimmermannswinkel

418 **carriage bolt**
[iwa/scr]
f boulon à collet
e tornillo de cuello
i bullone con colletto
n slotschroef
d Bundbolzen; Schlossschraube

419 **carriagebuilder's plane; carriagemaker's plane**
[pla]
f rabot de charron
e cepillo de carrocero
i pialla di carradore
n wagenmakersschaaf
d Wagenbauerhobel

420 **car tool set**
[too/tra]
f jeu d'outils de voiture
e caja de utensilios de coche
i borsa attrezzi per la macchina
n autoset
d Wagengerätekasten

421 **cartridge-operated hammer; nail gun**
[ham]
f canon pour clous
e cañon para clavos
i cannone per chiodi
n spijkerkanon
d Nagelkanone

422 **carving chisel**
[art/chi/wool]
f ciseau pour sculpteur en bois
e cincel de escultor en madera
i scalpello di scultore in legno
n beeldhouwersbeitel *(hout)*
d Bildhauerbeitel *(Holz)*

423 **carving gouge**
[art/chi/wool]
f gouge pour sculpteur en bois
e gubia de escultor en madera
i sgorbia di scultore in legno
n beeldhouwersguts *(hout)*; houtsnijguts
d Bildhauerhohlbeitel *(Holz)*

424 **car washing brush**
[cle/tra]
f brosse à laver les voitures
e cepillo lavador de coches
i spazzola lavatrice per macchine
n autowasborstel
d Wagenwaschbürste

425 **case opener; wrecking bar**
[eng]
f ciseau à déballer
e pie de cabra
i piede di porco
n kistbeitel
d Kistenbeitel

426 **case rack; case stand**
[cpt/pri]

f casse de lettres
e chibalete
i bancone da caratteri; bancone del compositore
n letterkast; zetkast
d Satzregal

427 **cask bottom hammer**
[coo/ham]
f marteau de fond de tonneau
e martillo de fondo de tonel
i martello da fondo di botte
n vatbodemhamer
d Fassbodenhammer

428 **castellated nut; castle nut**
[iwa/scr]
f écrou à entailles; écrou crénelé
e tuerca entallada
i dado a corona; dado ad intagli
n kroonmoer
d Kronenmutter

429 **casting ladle; foundry ladle**
[fou]
f cuiller à couler
e caldero de colada; cazo de colada
i cucchiaione da fonderia
n gietpan
d Giesskelle; Giesslöffel; Giesspfanne

430 **castings brush**
[fou]
f brosse pour pièces coulées
e escoba para piezas coladas
i spazzola per getti
n gietstukkenborstel
d Gussstückenbürste

* **castle nut** → **428**

431 **catch bolt**
[hou/iwa]
f verrou à ressort
e pestillo de resorte; picaporte de resorte
i scatto a molla
n veergrendel
d Federriegel

432 **ca(u)lking**
[ham/nav]
f calfat; calfatage
e calafateo
i calafatura
n breeuwen; kalfatering
d Kalfatern

433 **ca(u)lking; cocking; cogging; corking**
[car]
f matage
e mortaja
i incavatura
n aanzetten
d Stemmen

434 **ca(u)lking chisel**
[chi/nav]
f ciseau à mater; matoir
e calafateador; cincel de calafateo
i scalpello da presellatura; scalpello di calafato
n breeuwbeitel; kookbeitel
d Abdichtmeissel; Stemmeissel

435 **ca(u)lking gun; mastic gun**
[gen/gla/mas]
f pistolet de mastic
e pistolete de mástico
i pistola a spruzzo di mastice
n kitspuit
d Kittspritze

436 **ca(u)lking hammer**
[ham/nav]
f marteau de calfat
e bandarria; mandarria; martillo de calafateo
i martello da calafatura; martello di calafato
n breeuwhamer; kalfaathamer
d Kalfathammer

437 **ca(u)lking hammer; cogging hammer**
[car]
f marteau de matage
e martillo de mortaja
i martello da incavatura
n aanzethamer
d Stemmhammer

438 **ca(u)lking iron**
[ham/nav]
f ciseau de calfat
e hierro de calafatea
i ferro da calafatura; ferro di calafato
n kalfaatijzer
d Kalfateisen

439 **ca(u)lking iron; cogging iron**
[car]
f ciseau de matage
e hierro de mortaja
i ferro da incavatura
n aanzetijzer
d Stemmeisen

440 **ca(u)lking iron; ca(u)lking tool**
[cle/met]

f ciseau de radoub; clameau
e útil para tapar
i utensile per accecare
n breeuwijzer; kauwijzer
d Stopfeisen

441 cavil; jedding axe
[axe/ham]
f hache fendeuse; merlin
e hacha de rajar; martillo de hendar
i ascia fenditrice; martello fenditore
n kloofbijl
d Spaltaxt; Spalthammer

442 cavity-wall plug; hollow-wall plug
[bui/fas/iwa]
f bouchon de mur creux; tapon de mur creux
e tapón para muro sordo; tapón para pared hueca
i tapone per muro cavo; tapone per muro di tramezzo
n hollewandplug; spouwmuurplug
d Hohlmauerdübel

443 C clamp; C cramp; G clamp; G cramp
[cla]
f presse à vis en C
e prensa en C; sargento en C
i morsetto a C
n ketelklem
d Schraubzwinge

444 ceiling mop; pope's head; Turk's head
[cle/hou]
f tête de loup
e escobón
i granatella
n raagbol; ragebol
d Spinnwebfeger

445 cement chisel
[bui/chi]
f ciseau pour insérer le ciment
e cincel para insertar el cemento
i scalpello per inserire il cemento
n cementbeitel
d Zementmeissel

446 cement gun
[bui/gen]
f guniteuse
e pistola para enlucidos de cemento
i pistola spruzza cemento
n cementkanon; cementspuit
d Zementkanone

447 center(re) bit
[dri]
f mèche anglaise; mèche à trois pointes
e barrena de pala; broca de centrar
i mecchia a tortiglione; punta francese
n centerboor
d Zentrumsbohrer

448 center(re) drill; spotting drill
[dri]
f foret à centrer
e broca de centrar
i punta a centrare
n centreerboor
d Zentrierbohrer

449 center(re) finding ga(u)ge; center(re) square; radial square
[mea]
f équerre de centrage
e escuadra de centrar; escuadra de diámetros
i squadra di centraggio
n centreerhaak; hartzoeker
d Zentrierwinkel

450 center(re) hole
[dri]
f forage de centrage
e taladrado de centrar
i trapanatura di centratura
n centergat
d Zentrierbohrung

451 cent(e)ring drilling machine
[dri]
f perceuse à centrer
e taladro centrador
i trapano da centratura
n centreerboormachine
d Zentrierbohrmaschine

452 cent(e)ring pin
[iwa/mea]
f cheville ouvrière; goujon de centrage; tige-guide
e espárrago de centrar
i perno da centratura
n centreerpen
d Zentrierstift

453 center(re)less grinding
[gri]
f rectification sans pointes
e rectificado sin puntas
i rettifica senza centri
n puntloos slijpen
d spitzenloses Schleifen

454 center(re) punch
[per]

f perçoir; pointeau
e granete; punzón de marcar
i punzone per centri
n centerpons; keur; kornagel
d Ankörner; Körner; Locheisen

455 center(re) reamer
[rea]
f alésoir à centrer
e escariador centrador; escariador de centrar
i alesatore a centrare
n centreerruimer
d Zentrierreibahle

*** center(re) square → 449**

456 ceramic-bonded abrasive
[gri/san]
f abrasif à agglomérant céramique
e abrasivo ligado con cerámica
i abrasivo ad impasto di ceramica; abrasivo collato con ceramica
n keramisch gebonden slijpmiddel
d keramisch gebundenes Schleifmittel

457 chain block
[tra]
f poulie à chaine
e polea de cadena
i paranco a catena
n kettingtakelblok
d Kettenrolle

458 chain pipe wrench
[spa]
f pince à tuyaux à chaîne; serre-joint à chaîne
e llave de cadena para tubos
i chiave a catena per tubi; giratubi a catena
n kettingpijptang
d Kettenrohrzange

459 chain rivet extractor
[riv]
f chasse-rivets à chaine
e embutidor de remaches con cadena
i estrattore per chiodi a catena
n klinknageltrekker met ketting
d Nietenzieher mit Kette

460 chain saw
[saw]
f scie à chaîne
e sierra de cadena
i sega a catena
n kettingzaag
d Kettensäge

*** chairmaker's saw → 230**

461 chairman's hammer; gavel
[ham]
f marteau du président
e mazo presidencial
i mazzuola del presidente
n voorzittershamer
d Hammer eines Vorsitzenden

462 chalk line
[mas]
f cordeau au charbon; cordeau blanchi à la craie
e cuerda al carbón
i corda al carbone
n metselkoord; slaglijn; smetdraad
d Mauerschnur; Schlagschnur; Zimmerschnur

463 chamfer
[car/fre/woo]
f biseau
e bisel
i bisellatura; smussatura
n faas
d Fase

464 chamfer bit; chamfer cut
[car/fre/woo]
f fraise à biseauter les bords
e fresa para biselar contos
i fresa per bisellare bordi
n faasfrees; kantbreekfrees
d Fasenfräser; Kantenfräser

465 chamfering
[san/woo]
f exécution de biseaux
e biselado
i bisellatura; smussatura
n afschuinen van randen
d Fasen

466 chamfering hammer
[car/ham]
f marteau à biseauter; marteau à chanfreiner
e martillo biselador
i martello di bisellatura; martello per smussare
n afschuinhamer
d Abschräghammer

467 chamfer plane
[car/pla]
f rabot à chanfreiner
e guillame de inglete
i pialletto a smussatura

n vellingschaaf
d Schräghobel

468 chamfer spokeshave
[sha]
f vastringue à chanfreiner
e bastren de cantear
i pialletto per sagomare a smussatura
n vellingspookschaaf
d Kantenschabhobel

469 channel cutter; groove cutter
[lea]
f traceret
e cuchillo acanalador
i lesina
n ritsmes
d Rissmesser

470 Charpy hammer
[ham]
f marteau de Charpy
e martillo de Charpy
i martello di Charpy
n charpyhamer; kerfslaghamer
d Charpy-Hammer

471 charring hammer; scutch hammer; stone dressing hammer
[ham/sto]
f boucharde
e bujarda
i bocciarda
n kareerhamer; schareerhamer
d Charierhammer; Quarrierhammer; Scharrierhammer

472 chaser
[fre/scr]
f mâchoire à peigne
e cabezal de roscar; peine de roscar
i fresa per filettare; pettine per filettare
n schroefdraadfrees
d Schneidbacken

* **chasing chisel → 400**

473 chasing cutter; groove bit
[cpt/fre]
f fraise à rainurer
e fresa para ranurar
i fresa per scanalare
n groeffreeskop
d Nutenfräser; Rillenfräser

474 chasing hammer
[ham]
f marteau à repousser
e martillo para abollonar; martillo para repujar
i martello per stozzare
n drijfhamer
d Treibhammer

475 chasing hammer; enchasing hammer
[dec/ham]
f marteau à ciseler
e martillo para cinclar
i martello da cesello
n ciseleerhamer
d Ziselierhammer

476 chasing saw
[saw]
f scie pour faire des rainures dans les murs
e sierra para hacer ranuras en muros
i sega per fare scanalature nei muri
n zaag voor het aanbrengen van sleuven in muren
d Säge zum Ziehen von Nuten in Mauern

477 check nut; lock nut
[fas/iwa/scr]
f contre-écrou
e contratuerca
i controdado
n borgmoer; contramoer
d Kontermutter

478 cheeks
[cpt/saw]
f joues
e gualderas tensoras
i traverse
n raamhoofden
d Wangen

479 cheese head screw
[fas/iwa/scr]
f vis à tête ronde
e tornillo de cabeza redonda
i vite a testa semisferica
n bolkopschroef; schroef met bolle kop
d Rundkopfschraube

480 chime plane; rim plane
[coo]
f rabot de bord
e cepillo de borde
i pialla di bordo
n kimschaaf
d Kimmenhobel; Randhobel; Zargenhobel

481 chimney sweeper's brush with weight
[cle/hou]
f hérisson avec boulot
e deshollinador

i spazzola con peso
n borstel met kogel
d Sonne mit Kugelschlag

482 chip-axe
[foy/woo]
f épaule de mouton
e azuela; doladera
i ascia da digrossare
n slechtbijl
d Schlichtbeil

483 chipping chisel; cold chisel
[chi/met]
f ciseau à froid
e cortafrío
i scalpello a freddo; scalpello da banco
n koudbeitel
d Hartmeissel; Kaltmeissel

484 chipping hammer; welder's chipping hammer
[ham/wel]
f marteau à piquer les soudures; marteau à scories
e martillo de desbastar; martillo para picar soldaduras
i martellina battiscorie; martello per l'asportazione delle scorie; martello di saldatore
n lasbikhamer; slakkenhamer
d Schlackenhammer; Schweisserhammer

485 chipping knife
[for]
f bouloir de maréchal; ébarboir; racloir
e cuchilla desbarbadora
i coltello da raschiare
n schraapmes
d Abgratmesser; Kratzmesser

486 chisel blade
[chi/cpt]
f lame de ciseau
e hoja de cincel
i lamina di scalpello
n beitelblad
d Beitelblatt

487 chisel bolster
[chi]
f épaulement de ciseau
e hombro de cincel
i spallamento di scalpello
n borst van een beitel
d Scheibe zwischen Angel und Klinge

488 chisel collar
[chi/cpt]
f col de ciseau
e collar de cincel
i collare di scalpello
n beitelkraag
d Beitelkragen

489 chisel for inside work
[chi/met/woo]
f ciseau pour tournage intérieur
e cincel para trabajos interiores
i scalpello per lavori interni
n binnenbeitel
d Innendrehbeitel

490 chisel handle
[chi/cpt]
f manche de ciseau
e mango de cincel
i manico di scalpello
n beitelhecht
d Beitelheft

491 chisel holding tongs
[chi/pli]
f tenaille(s) à burins
e tenazas de retención de cinceles
i tenaglie da ritenuto di scalpelli
n beiteltang
d Beitelzange

492 chisel neck; neck of a chisel
[chi/cpt]
f collet de ciseau
e cuello de escoplo
i collo di scalpello
n beitelhals; hals van een beitel
d Beitelhals

493 chisel rest
[chi/cpt]
f support de ciseau
e soporte de cincel
i supporto di scalpello
n leunspaan
d Beitelstütze

494 chisels
[chi]
f ciseaux
e cinceles; escoplos
i scalpelli
n beitels
d Beitel; Meissel

495 chisels for masonry works
[chi/sto]
f ciseaux du sculpteur en pierre
e cinceles de escultor en piedra
i scalpelli di scultore in pietra

n beeldhouwersbeitels *(steen)*
d Bildhauermeissel *(Stein)*

496 chisel sharpening template
[chi/cpt]
f gabarit d'affûtage pour ciseaux
e plantilla rectificadora para cinceles
i sagoma rettificatrice per scalpelli
n beitelslijpmal
d Schleifschablone für Beitel

497 C hook spanner; hook spanner
[spa]
f clé à ergot plat
e llave de gancho; llave de tetones
i chiave a settore
n haaksleutel
d Hakenschlüssel

498 chop; jaw
[cla/cpt/too]
f mâchoire
e mandíbula
i morsetto da banco
n wang
d Backe

499 chopper; cleaver
[agr/foy/hor]
f couperet
e tajadera
i mezzaluna; pestarola
n hakmes
d Hackmesser

500 chopper knife
[agr]
f hache-paille; hacheuse
e cuchilla cortapajas
i coltello tagliuzzatore
n hakselmes
d Häckselmesser

501 chopping board
[nut]
f billot
e tajo
i banco da macellaio
n hakblok
d Hackbank

502 chopping hoe
[agr/hor]
f houe
e azadón de rozar
i zappa
n hak
d Schlaghacke

503 chuck
[cpt/dri]
f mandrin à serrage
e mandril; portabroca
i mandrino; portapunta
n boorhouder; klauwplaat
d Klemmfutter; Spannfutter

504 chuck adapter
[cpt/dri]
f adaptateur pour mandrin
e adaptador para mandril
i adattatore per mandrino
n aanpasstuk voor boorhouder
d Adapter für Bohrfutter

505 chuck guard
[cpt/dri]
f protection de mandrin
e protector de mandril
i protezione di mandrino
n boorhouderbeveiliging
d Bohrfutterschutz

506 chuck jaws
[cpt/dri]
f becs de mandrin
e mordazas de mandril
i ganasci di mandrino
n boorhouderbekken
d Bohrfutterbacken

507 chuck key
[cpt/dri]
f clé pour mandrin
e llave de mandril
i chiave di mandrino
n boorhoudersleutel
d Bohrfutterschlüssel

508 chuck nut
[cpt/dri]
f écrou de mandrin
e tuerca de mandril
i dado di mandrino
n boorhoudermoer
d Bohrfuttermutter

509 chuck type tap wrench
[cpt/dri]
f porte-tarauds
e pequeña terraja de cojinete
i giramaschio piccolo
n tapkruk
d Gewindebohrerhalter; kleines Windeisen

* **churn drill → 390**

510 circle cutter
[cut/dri]
f coupe-cercle
e cortador de círculo
i tagliatore di cerchio
n cirkelsnijder; cirkeltrekker
d Kreisschneider

511 circle sprinkler
[hor]
f arroseur rotatif
e rociadero de círculo
i girandola; irroratrice
n cirkelsproeier
d Kreissprinkler

512 circlip pliers
[pli]
f pince pour circlips
e alicate(s) para arandelas de seguridad
i pinza per anelli di sicurezza
n borgringtang
d Sicherungsringzangen

513 circular grooving saw
[saw]
f scie circulaire à rainurer
e sierra circular para ranurar
i sega circolare per scanalare
n sponningcirkelzaag
d Nutkreissäge

514 circular level
[mea]
f niveau circulaire
e nivel circular
i livella circolare
n doosniveau; dooswaterpas; luchtbelwaterpas
d Dosenlibelle

515 circular plane; compass plane
[pla]
f rabot flexible; rabot cintrable
e cepillo para superficies curvas; rebajador
i sponderuola a barca
n hol- en bolschaaf; toogschaaf
d offener Grundhobel; Schiffshobel (verstellbar)

516 circular rebate plane; compass rebate plane
[car/pla/woo]
f guillaume rond
e guillame rondo
i sponderuola a canto di gattuccio
n ronde rabatschaaf
d Rundfalzhobel

517 circular saw
[saw]
f scie circulaire
e sierra circular
i sega circolare
n cirkelzaag
d Kreissäge

* **circular saw** *GB* → **382**

* **circular saw axis** *GB* → **383**

* **circular saw bank** *GB* → **384**

518 circular saw blade
[cpt/saw]
f lame de scie circulaire
e hoja de sierra circular
i lama di sega circolare
n cirkelzaagblad
d Kreissägeblatt

* **circular saw blade** *GB* → **385**

519 circular saw blade with straight teeth
[cpt/saw]
f lame de scie circulaire à denture droite
e hoja de sierra circular de dentadura recta
i lama di sega circolare a dentatura retta
n cirkelzaagblad met rechte vertanding
d Metallkreissägeblatt mit Normalverzahnung

* **circular saw file** *GB* → **386**

520 circular solid die
[cut/met/scr]
f filière ronde solide
e cojinete de terrajar sólido
i filiera tonda solida
n ronde snijplaat gesloten model
d geschlossenes Gewindeschneideisen

521 circular split die
[cut/met/scr]
f filière ronde extensible
e cojinete de terrajar extensible
i filiera tonda regolabile
n ronde snijplaat gespleten model
d geschlitztes Gewindeschneideisen

522 circumference rule
[mea]
f règle pour mesurer le contour
e regla medidora de contorno
i regolo misuratore di contorno
n omtrekmeetlat
d Umrissmessstab

523 **circumference scale**
[meal]
f échelle de contour
e escala de contorno
i scala di contorno
n omtrekschaal
d Umrissskale

524 **clamp; cramp**
[car/cla/gen]
f sergent; serre-joint
e cierrajuntas; sargento
i sergente; strettoio
n klem
d Zwinge

525 **clamped miter(re) box**
[cpt/saw]
f boîte à onglets à serrage
e caja para cortar en inglete de enclavación
i cassetta per cornici a serraggio
n klemmende verstekbak; klemmende versteklade
d Gehrungslade mit Klemmeinrichtung; Schneidelade mit Klemmvorrichtung

526 **clamp heads; cramp heads**
[cla/cpt]
f mâchoires de serrage
e mandíbulas; mordazas
i ganasce di serraggio
n klembekken; klemkoppen
d Klemmbacken

527 **clapper; clapper tool box; tool clapper**
[mec/tool]
f porte-outil basculant; porte-outil de mortaiseuse horizontale
e portaherramienta basculante; portaherramienta de limadora horizontal
i portautensile basculante; portautensile per limatrice orizzontale
n verende beitelhouder
d Meisselklapper; Messerhalterklapper

528 **claw; tack puller**
[car/cpt/woo]
f arrache-clous
e garra sacaclavos; pie de cabra
i pinza per chiodi
n spijkerklauw; spijkertrekker
d Geissfuss; Nagelklaue; Nagelzieher

529 **claw chisel; tooth chisel**
[chi/sto]
f gradine à pierre
e estique
i gradina
n tandijzer
d Zahneisenmeissel

530 **claw hammer**
[car/ham]
f marteau à dent; marteau fendu; marteau tire-clou
e martillo de orejas
i martello a granchio; martello con testa a granchio
n klauwhamer
d Klauenhammer

531 **claw spanner**
[spa]
f clé à griffe
e llave de oreja
i chiave a piede di gallo
n hanepootsleutel; klauwsleutel
d Stirnlochschlüssel

532 **cleaner**
[cle/dec]
f racloir
e rascador
i raschiatoio; raschietto
n schraper; schrapper
d Kratzer

533 **cleaning needle**
[bur/cle/cpt]
f aiguille à nettoyage
e aguja de limpieza
i ago da pulitura
n doorsteeknaald
d Durchstecknadel

534 **clearing the cutting path**
[saw]
f dégageage du tracé
e limpieza de la línea de corte
i nettare della linea di taglio
n schoonhouden van de zaaglijn
d Reinhaltung der Schnittfuge

535 **cleaver; river; splitting knife**
[coo]
f fendoir
e rajador
i fenditoio
n kloofmes
d Klobeisen; Spaltklinge

* **cleaver → 499**

536 **cleaving chisel**
[chi/sto]
f rabattoir
e cincel de hender

i scalpello da spaccamento
n kliefbeitel; kloofbeitel
d Spaltmeissel

537 **cleaving saw; pit saw; ripper saw; rip saw**
[foy/saw]
f scie à refendre
e sierra de cortar al hilo; sierra de cortar a lo largo; sierra de hilo
i sega per il taglio in lunghezza
n schulpzaag
d Brettsäge; Klobsäge; Längensäge; Spaltsäge

* **clip → 303**

538 **cloth disk**
[cpt/gri/san]
f disque à toile
e disco de tela de esmeril
i disco a tela smerigliata; disco a tela vetrata
n slijpschijf met schuurlinnen
d Leinenschleifscheibe

539 **clout nail**
[iwa]
f caboche
e clavo para el papel asfaltado; tachuela
i chiodo a testa larga; chiodo per cartone castramato
n asfaltnagel; asfaltspijker
d Pappnagel

540 **club hammer; hand drilling hammer; lump hammer**
[ham]
f masse
e maceta
i mazzetta
n vuist; vuistmoker
d Fäustel

541 **club hammer with bevelled edges**
[cpt/ham]
f masse carrée à angles abattus; masse carrée à angles chanfreinés
e maceta de bordes biselados
i mazzetta a bordi smussati
n vuist met afgeschuinde kanten
d Fäuste mit gebrochenen Kanten

542 **club hammer with square edges**
[cpt/ham]
f masse carrée à angles vifs
e maceta de bordes rectangulares
i mazzetta a bordi rettangolari
n vuist met scherpe kanten
d Fäuste mit rechteckigen Kanten

543 **coach screw**
[iwa]
f tire-fond
e tirafonda
i vite a mordente con testa squadrata
n slotschroef
d Stehbolzen

544 **coal shovel**
[gen/hou]
f pelle à charbon
e pala para carbón
i pala di carbone
n kolenschop
d Kohlenschaufel

545 **coarse cut**
[far]
f taille rude
e picadura gruesa
i taglio grosso
n grove hak
d Grobhieb; Hieb O

* **coarse file → 84**

546 **coarse hackle**
[agr/tex]
f ébauchoir; gros peigne; séran grossier
e peine en grueso; rastrillo en grueso
i pettine da sgrossatura; pettine grosso; pettine per la prima pulitura
n aftrekhekel
d Abzughechel; Grobhechel

547 **coarseness of cut**
[far]
f degré de rudesse de taille
e grado de grueso de picadura
i grado di grossezza di taglio
n grofte van kap
d Hiebgrobheit

548 **coarse toothing**
[saw]
f denture grosse
e dentadura gruesa
i dentatura grossa
n grove vertanding
d Grobverzahnung

549 **cobbler's awl**
[lea]
f alène de cordonnier
e lezna de zapatero
i lesina di calzolaio

n schoenmakersels
d Vorstecher

550 cobbler's knife; shoe knife
[lea]
f tranchet
e tranchete
i trincetto
n schoenmakersmes
d Kneif; Schustermesser

551 cobbler's rasp; shoe rasp
[far/lea]
f râpe de bottier; râpe de cordonnier
e escofina de zapatero
i raspa di calzolaio
n schoenmakersrasp
d Schuhmacherraspel; Schusterraspel

*** cocking → 433**

*** cogging → 433**

*** cogging hammer → 437**

*** cogging iron → 439**

552 cog wheel puller; toothed wheel puller
[gen/pull]
f extracteur de roue dentée
e extractor de rueda dentada
i estrattore di ruota dentata
n tandwieltrekker
d Zahnradabzieher

553 colander; cullender
[cle/hou]
f passoire
e pasador
i colabrodo; colatoio
n vergiet
d Durchschlag

*** cold chisel → 483**

554 cold metal saw
[saw]
f scie froide à métaux
e sierra para metales en frío
i sega da metallo a freddo
n koude metaalzaag
d Kaltmetallsäge; Metallsäge zum kalten Schneiden von Metall

555 collapsible handle
[cpt/saw]
f manche fermant
e mango rebatible
i manico rientrante
n inklapbaar handvat
d Klappheft

556 collar
[agr/hor/gen]
f frette; virole
e virola
i ghiera; virola
n dil; dille; steelhuis
d Frette; Zwinge

*** collar → 11**

557 collar grip
[cpt/dri]
f poignée-collier
e empuñadura de collar
i impugnatura a collare
n klemzijhandgreep
d Schellengriff

558 collet
[cut]
f douille de serrage
e manguito de acoplamiento
i bussola
n collet; klemhuls; klemring
d Klemmbüchse; Spannbüchse

559 collet chuck
[cut]
f mandrin de serrage
e mandril de anguito
i mandrino a bussola
n collethouder
d Büchsenfutter

560 comb; graining comb; marbling comb
[dec]
f peigne à décor; peigne de peintre
e peine jaspeador
i pettine per marmorizzare
n marmerkam
d Marmorierkamm

561 combination bevel
[mea]
f équerre fausse composée
e falsa escuadra combinada
i squadra falsa combinata
n samengestelde zwei
d Kombinationsschmiege

562 combination bits; combination cutters
[car/fre/woo]
f fraises combinées
e fresas combinadas
i frese combinate

n combinatiefrezen
d Kombinationsfräser

563 **combination gas pliers**
[pli]
f pinces universelles à gaz
e alicate(s) universal(es) de gasista
i pinze universali per tubi
n universele gaspijptangen
d Allzweckgasrohrzangen

564 **combination plane**
[pla]
f rabot à combinaisons; rabot monobloc
e cepillo de combinación
i pialla combinata
n combinatieschaaf
d Kombinationshobel

565 **combination pliers**
[pli]
f pinces universelles
e alicates universales
i pinze universali
n combinatietangen
d Kombinationszangen; Kombizangen

566 **combination saw blade**
[cpt/saw]
f lame de scie à dentures combinées
e hoja de sierra de dentaduras combinadas
i lama di sega con combinazione di denti
n combinatiezaagblad
d Sägeblatt mit kombinierter Zahnung

567 **combination set**
[mea]
f équerre universelle à limbe
e escuadra universal con transportador
i squadra universale con rapportatore
n combinatiehaak met gradenboog
d Universalwinkelmesser mit Gradbogen

568 **combination sharpening stones**
[gri/sha]
f pierres à affûter jumelées
e piedras combinadas de amolar
i pietre combinate da affilare
n gecombineerde banksteen; tweelaagsaanzetsteen
d Kombinationsschärfsteine

569 **combination spanner**
[spa]
f clé anglaise à anneau; clé mixte
e llave inglesa con boca; àve mezclada
i chiave combinato; chiave inglese ad anello
n ring-steeksleutel; steek-ringsleutel
d Gabel-Ringschlüssel; Ring-Gabelschlüssel

570 **combination spokeshave**
[pla]
f vastringue à combinaisons
e bastren de combinación
i pialletto combinato per sagomare
n combinatiespookschaaf
d Kombinationsschabhobel

571 **combination square**
[mea]
f équerre universelle
e escuadra universal
i squadra universale
n combinatiehaak
d Universalwinkelmesser

572 **combined hoe and fork; double hoe**
[agr/hor]
f serfouette
e combinada de azado y horquilla
i zappetta a doppio uso
n schoffelhark; tuinhakje
d Dreispitz; Kombi-Hacke

573 **combined rivet set and snap**
[riv]
f butterolle en combinaison
e uñeta y buterola combinada
i chiodaia e stampo per ribattini combinato
n gecombineerde ophaler en dopper
d Nietenquetscher-Kombination

574 **comb-maker's saw**
[saw]
f scie de peigneran; scie de peignier; scie double
e sierra de peinero; sierra doble
i sega da pettinaio; sega doppia
n raspzaag; tandzaag
d Kammacher's Säge

575 **compass; magnetic compass**
[mea/nav/sur]
f boussole
e brújula
i bussola
n kompas
d Bussole; Kompass

* **compass plane** → **515**

* **compass rebate plane** → **516**

576 **compass saw; lock saw; wallboard saw**
[saw]

f scie à guichet
e sierra de punta; sierra de puñal
i sega a foretto
n scherpzaag; schrobzaag; stootzaag
d Stichsäge

577 **compass saw blade**
[cpt/saw]
f lame guichet
e hoja de sierra de punta; hoja de sierra de puñal
i lama di sega a foretto
n schrobzaagblad
d Stichsägeblatt

578 **compound press tool**
[tool]
f matrice composée pour presses
e estampa compuesta; traquel compuesto de prensa
i matrice composta di pressa
n samengestelde persmatrijs
d Verbundpressmatrize

* **compressed air hammer → 38**

579 **concave cutter**
[fre]
f fraise profil concave
e fresa perfil cóncava
i fresa profilo concavo
n concaaffrees
d Halbkreisformfräser nach innen gewölbt (konkav)

580 **concave jointer**
[mas]
f fer de joint concave
e paletín cóncavo
i ferro concavo per giunti di mattoni
n holle voegspijker
d Hohlfugenkelle

581 **concrete drill bit**
[bui/dri]
f foret pour béton
e broca para hormigón
i punta da trapano per calcestruzzo
n betonboor
d Betonbohrer

582 **concrete nail; steel nail for concrete**
[iwa]
f clou pour béton
e clavo para hormigón
i chiodo per calcestruzzo
n betonduim; betonnagel
d Betonnagel

583 **concrete point; pointed chisel**
[chi]
f ciseau pointu
e puntero
i scalpello appuntito; subbia
n puntbeitel
d Spitzmeissel

584 **concreter's nippers; top cutting pliers**
[pli]
f pince Monier; tenaille(s) russe(s)
e pinza para hormigón armado; tenaza rusa
i tenaglie da calcestruzzo
n draadvlechttang; moniertang
d Monierzange

585 **concreter's tongs**
[pli]
f tenaille(s) pour bétonneurs
e cortaenrejado
i tenaglie di cementista
n betonwerkerstang
d Rabitzzange

586 **concrete scraper**
[bui/cpt]
f lame à béton
e rasqueta para hormigón
i raschietto per calcestruzzo
n betonschraper
d Betonklinge

587 **concreting trowel; flooring trowel**
[mas]
f plâtroir pour sols; truelle à béton
e paleta para hormigón; palustre para suelos
i cazzuola per calcestruzzo; frettazzo per pavimenti
n betonplakspaan; vloerplakspaan; vloerspaan
d Betonkelle; Bodenlegerkelle

588 **cone-cut sheet drill**
[dri]
f foret à rectification conique pour tôles
e broca de rectificación cónica para hojas
i punta a rettifica conica per lamiere
n konisch geslepen boor voor blik
d konisch geschliffener Blechbohrer

589 **cone-cut tube drill**
[dri]
f foret à rectification conique pour tubes
e broca de rectificación cónica para tubos
i punta a rettifica conica per tubi
n konisch geslepen boor voor buizen
d konisch geschliffener Bohrer für Röhren

590 cone pliers
[pli]
f pince à cone
e tenaza de cono
i tenaglie a cono
n conustang
d Konuszange

591 conical countersink
[dri/rea]
f alésoir-fraise conique
e avellanador cónico
i accecatoio conico
n konische verzinkfrees; soevereinboor
d Kegelsenker

592 conical reamer
[dri/fre/rea]
f alésoir conique
e escariador cónico
i alesatore conico
n konische ruimer
d Kegelreibahle

593 conical spike
[gen/iwa]
f goupille conique
e pasador cónico
i caviglia conica
n konische stift
d Kegelstift

594 conical threaded detachable shaft
[cpt/gri/san]
f arbre amovible conique
e árbol amovible cónico
i albero smontabile conico
n konische afneembare spil
d konische abnehmbare Welle

595 connecting bolt; fastening bolt
[iwa]
f boulon d'assemblage
e perno de unión
i bullone di fissaggio
n aanzetbout; verbindingsbout
d Verbindungsbolzen

596 connecting shaft
[cpt]
f bielle
e biela
i biella
n drijfstang
d Pleuel

597 connecting spindles
[cpt]
f arbres de connexion
e husillos de conexión
i alberi di connessione
n aansluitasjes
d Anschlussspindel

598 constructional spanner; offset spanner
[spa]
f clé contrecoudée
e llave acodada
i chiave curva
n gebogen ringsleutel; opsteeksleutel
d Aufsteckringschlüssel

599 contact file
[ele/far]
f lime contact
e lima de contacto
i lima per contatti
n contactpuntvijl; contactvijl
d Kontaktfeile

600 continuous hinge
[iwa]
f charnière continue
e bisagra continua
i cerniera a metraggio
n bandscharnier
d Bandscharnier; Serienscharnier

601 contraction rule; patternmaker's rule
[car/fou/mea]
f mètre à retrait
e metro de contracción; metro de modelista
i metro con ritiro; metro di modellista
n krimpduimstok; modelmakersduimstok
d Schwindmasstab

602 contractor's tools
[gen/too]
f outils d'entrepreneur
e herramientas de contratista de obras
i utensileria di appaltatore
n aannemersgereedschap
d Bauunternehmerwerkzeuge

603 convex cutter; convex miller
[fre]
f fraise gauchette; fraise profil convexe
e fresa perfil cónvexo
i fresa profilo convesso
n convexfrees
d Halbkreisformfräser nach aussen gewölbt (konvex); Hohlfräser

604 cooper
[coo]
f tonnelier
e cubero; tonelero
i barilaio; bottaio

n kuiper
d Böttcher; Küfer

605 cooper's bench; shaving horse
[coo]
f bastringue; chevalet de tonnelier; marotte
e mochuelo
i banco di taglio per le doghe
n reepbank; strijkbank
d Schneidbank; Schnitzelbank

606 cooper's chisel
[chi/coo]
f clouet
e cincel de cubero; cincel de tonelero
i scalpello di barilaio; scalpello di bottaio
n kuipersbeiteltje
d Böttcherbeitel; Küferbeitel

607 cooper's knife; drawing knife; hollowing knife
[coo]
f cochoir; débordoir; plane cintrée
e cuchillo de tonelero; plana arqueda
i coltello a petto a lama ricurva
n baars; haalmes; holmes; steekmes
d Fasszieher; Krummeisen; Rundschaber

608 cooper's plane; round shave
[coo]
f colombe; plane à genoux; rabot de tonnelier
e cepillo redondeador
i pialla di barilaio; pialla di bottaio
n kuipersschaaf
d Böttcherschaber; Küferschaber

609 cooper's set hammer
[coo/ham]
f chasse de tonnelier
e asentador de tonelero; destajador de tonelero
i spiana per rinforzare i cerchi di ferro
n kuiperszethamer
d Küfersetzhammer

610 cooper's windlass
[coo]
f serre-joint; tiretoire; vérin de serrage
e gato de armar
i arganello a vite
n klemhaak; ringtang
d Fasszieher; Fasszug

611 coordinates bank
[dri/mea]
f banc de coordonnées
e banco de coordenadas
i banco di coordinate
n coördinatenbank
d Koordinatenbank

612 coordinates drilling machine
[dri]
f perceuse à coordonnées
e taladro de coordenadas
i trapano a coordinate
n coördinatenboormachine
d Koordinatenbohrmaschine

613 coping saw
[saw]
f scie à découper
e sierra de recortar
i sega da traforo
n decoupeerzaag; kromzaag; wipzaag
d Dekupiersäge

614 coppersmith's hammer
[ham]
f marteau de chaudronnier
e martillo de calderero en cobre
i martello di calderaio; martello di ramaio
n koperslagershamer
d Kupferschmiedehammer

615 core box bit; core box cutter
[car/fre/woo]
f fraise 1/2 sphérique
e fresa 1/2 esférica
i fresa 1/2 sferica
n halfronde holfrees
d Halbstabhohlfräser

616 core drill
[dri/min]
f trépan de sondage
e barrena hueca; sacatestigos
i carotiere
n buisboor; kernboor
d Kernbohrer

617 core drilling bit; diamond drilling bit
[dri/min]
f trépan à carotte
e taladro de corona con puntas de diamante
i perforatore a nucleo scorrevole
n beitelboor met diamantopleg; diamantboor
d Kernbohrgerät

*** corking → 433**

618 cork rubbing block
[pol]
f bloc de liège à poncer
e bloque de corcho para alisar

i blocco di sughero per lucidare
n schuurkurk
d Schleifkork

619 corner brace; gear frame brace
[dri]
f vilebrequin angulaire
e berbiquí angular
i menarola angolare
n hoekbooromslag
d Eckbohrwinde

620 corner chisel
[chi/woo]
f ciseau à tranchant angulaire
e cincel de ángulo
i scalpello ad angolo
n beitel met rechthoekige snede
d Winkelbeitel

621 corner clamp; corner cramp; edge clamp; edge cramp
[cla]
f serre-joint à coins
e sargento de bordes
i sergente ad angoli
n hoekklemschroef; kantklemschroef
d Hilfszwinge; Kantenzwinge

622 corner drilling machine
[dri]
f perceuse à coins
e taladro para esquinas
i trapano per angoli
n hoekenboormachine
d Eckbohrer

623 corner tool
[fou]
f lissoir angulaire
e alisador angular
i lisciatoio ad angolo
n haakse slikker
d Winkelpolierknopf; Winkelreibahle

624 corner trowel
[mas]
f truelle d'angle
e llana acodada; llana de ángulo; llana de esquinar
i cazzuola angolare
n hoektroffel
d Eckenkelle

625 cornice plane
[car/pla/woo]
f rabot à corniches
e cepillo para molduras a cornicadas
i sponderuola per cornici
n kornisschaaf; lijstschaaf
d Karnieshobel

626 corn shovel ***GB*****; grain shovel** ***US***
[agr/nut]
f pelle à remuer
e pala para apalear el grano
i pala da rivolgere il grano
n graanschop; korenschop
d Getreideschaufel

627 corrugated base; ribbed sole
[cpt/pla]
f semelle rainurée
e base acanalada
i base rigata
n geribde zool
d gerillte Sohle

628 corrugated sheet cutting
[cut/iwa]
f découpage de feuilles ondulées
e corte de chapa ondulada
i taglio di lamiera ondulata
n uitsnijden van golfplaat
d Ausschneiden von Wellblech

*** corundum paper → 49**

629 corundum wheel
[cpt/gri/san]
f meule à corindon
e muela de corindón
i mola a corindone
n korundslijpschijf
d Korundschleifscheibe

630 cotter pin
[fas]
f clavette à fourche; goupille d'arrêt
e chaveta de horquilla; pasador de aletas; pasador de chaveta
i copiglia
n borgspie; contraspie
d Vorstecker; Vorsteckstift

631 counter
[mea]
f compteur
e contador
i contatore
n telapparaat
d Zähler

632 counterbore cutter
[fre]
f fraise pour suraléser
e fresa escariadora
i fresa per accecatura

n verzinkfrees
d Versenkfräser

633 counter core
[fou]
f contre-noyau
e contramacho
i contro-anima
n contrakern; contraring
d Gegenkern

634 counter ga(u)ge; mortise ga(u)ge
[mea]
f trusquin
e gramil de doble punta; gramil de mortaja
i graffietto a due punte
n ritshout
d Zapfenlochlehre; Zapfenstreifmass

635 counter pattern
[car/mea]
f contre-gabarit
e contraplantilla
i controsagoma
n contramal; contrasjabloon
d Gegenschablone

636 counter pin
[fas]
f conte-clavette
e contrapasador
i controcopiglia
n contraspie
d Gegenstift

637 countersink
[car/cpt/dri/fre]
f alésoir-fraise; foret à fraiser
e avellanador
i accecatoio
n verzinker
d Senker; Versenker

638 countersink bit
[dri]
f foret à fraiser
e broca avellanadora
i punta di accecatura
n soevereinboor; verzinkboor
d Holzversenker

639 countersunk cutter
[dri/fre]
f fraise conique
e fresa cónica
i punta fresatrice
n schroefverzinkboor
d Versenkbohrer

640 cove bit; cove cutter
[car/fre/woo]
f fraise 1/4 concave
e fresa para moldura cóncava de cuarto de círculo
i fresa per profilare concava di quarto di circolo
n kwartrondholfrees
d Viertelstabhohlfräser

641 cove trowel
[mas]
f truelle à voûter
e llana para molduras cóncavas
i cazzuola per soffiti a ricasso
n ronde-hoekspaan
d Spiegeldeckenkelle

642 cradle
[bui/saf]
f échelle suspendue; échelle volante
e escala suspendida; escala volante
i scala sospesa; scala volante
n hangladder
d Hängeleiter

643 cramp
[cla/cpt/iwa]
f crampon de scellement
e grampillón
i arpione
n dookkram
d Ankereisen; Krampe

* **cramp → 524**

* **cramp heads → 526**

644 crampiron; staple
[cla/iwa]
f cavalier; crampon
e grapa
i cambretta; cavallottino
n draadkram
d Drahtklammer; Drahtkrampe

645 crane
[tra]
f grue
e grúa
i gru
n kraan
d Kran

646 crevice brush; radiator brush
[dec]
f brosse de radiateurs
e pincel para radiadores
i pennello per radiatori

n radiatorkwast
d Radiatorenbürste

647 crew punch; oblong punch
[lea]
f emporte-pièce à trous oblongs
e sacabocado(s) para agujeros alargados
i fustella per fori ablunghi
n holpijp voor langwerpige gaten; slagijzer
d Hohleisen für längliche Löcher

648 cross chisel; locksmith's chisel
[chi]
f ciseau à biseau cruciforme; ciseau de serrurier
e cincel de cerrajero; cincel de corte cruciforme
i scalpello a smusso crociforme; scalpello di magnano
n beitel met kruisvormige snede; slotenmakersbeitel
d Meissel mit kreuzförmiger Fase; Schlossermeissel

649 cross-cut; double cut
[far]
f à deux tailles; taille croisée; seconde taille
e picadura cruzada; picadura en cruz
i taglio doppio
n dubbele kap
d Kreuzhieb; Oberhieb

650 cross-cut chisel
[chi]
f ciseau à froid à queue carrée
e cortafrío con mango cuadrado
i scalpello a freddo a stelo quadro; unghietta
n koudbeitel met vierkante schacht; kruisbeitel
d Kreuzmeissel

* **cross-cut chisel → 400**

651 cross-cut saw
[saw]
f scie à tronçonner
e sierra tronzadora
i sega a traversa; sega per taglio trasversale
n afkortzaag
d Ablängsäge

652 cross-cut saw blade
[cpt/saw]
f lame de scie à tronçonner
e hoja de sierra tronzadora
i lama di sega a traversa; lama di sega per taglio trasversale
n afkortzaagblad
d Ablängsägeblatt

653 crossed half-lap joint; crossed halving
[car/joi/woo]
f assemblage croisé à mi-bois
e ensambladura cruzada a media madera
i giunto incrociato a mezzo legno
n halfhout-kruisverbinding
d Halbholzkreuzverbindung

654 cross filing
[far]
f limage croisé
e limadura cruzada
i limatura in croce
n kruiselings vijlen
d kreuzweises Feilen

655 cross joint
[car/joi/woo]
f assemblage en croix; jonction en croix
e ensambladura cruzada
i giunto incrociato
n kruisverbinding
d Kreuzverband; Kreuzverbindung

656 cross pein hammer
[ham]
f marteau à panne à travers
e martillo de peña transversal
i martello a penna trasversale
n bankhamer met dwarse pen; penbankhamer
d Bankhammer

657 crowbar; gablock; gavelock
[bui/gen]
f pied de biche
e palanca de pic de cabra; palanqueta
i palanchino
n breekijzer; koevoet
d Brecheisen; Brechstange

658 croze
[coo]
f jablière; jabloir; jabloire
e argallera; jabladera ruñador
i caprugginatoio
n gergelkam; gergelmes; kreusmes; kroosmes; sponmes
d Gargelhobel; Gargelkamm; Gargelreisser; Kröseleisen

659 croze plane
[coo/pla]
f rabot à jabler
e argallera; cepillo para ruñar jabladera

i pialla per capruggine
n beugelschaaf; kroosschaaf; sponschaaf
d Gargelhobel

660 **croze saw**
[coo/saw]
f scie à jabler
e gubiador
i sega per capruggine
n sponzaag
d Gargelsäge

* **crucible → 1590**

661 **crucible tongs**
[fou/pli]
f pince à creuset
e tenaza para el crisol
i pinza da crogiolo
n kroestang
d Tiegelzange

662 **crumbler; scrubber**
[agr/hor]
f herse émotteuse-niveleuse-affineuse
e rotocultivador
i coltivatore
n grondverkruimelaar
d Krümler

* **cullender → 553**

663 **cultivator** *GB*; **grubber** *US*
[agr/hor]
f griffe sarcleuse
e cultivador
i sarchiatore
n cultivator
d Grubber; Jätekralle; Wühleisen

664 **cupboard stand**
[gri/pol]
f socle-armoire
e bastidor en forma de caja
i sostegno in forma di armadio
n kastvormige standaard
d Ständerschrank

665 **cup-shaped wire brush**
[cpt/gen/gri]
f brosse-boisseau métallique
e brocha de taza cónica
i spazzola a tazza conica con filo in acciaio ondulato
n komstaaldraadborstel
d Stahldrahttopfbürste

666 **curry comb**
[agr/cle]
f étrille
e almohaza
i striglia
n roskam
d Striegel

667 **curved pruning knife**
[agr/hor]
f serpette
e navaja de jardiniero
i coltello a roncola
n snoeimes
d Asthippe; Gartenhippe; Gartenmesser; Hipper

668 **curved scissors**
[she]
f ciseaux courbes
e tijeras curvas
i forbici ricurve
n gebogen schaar
d gebogene Schere

669 **curved tooth cut**
[far]
f denture curviligne
e dentadura arqueada; dentadura curva
i dentatura ricurva
n gebogen kap; gebogen tanding
d Bogenverzahnung

670 **curving blade saw**
[saw]
f scie ventrée
e sierra de hoja convexa
i sega a lama convessa
n buikzaag
d Bauchsäge

671 **cut; file cut**
[far]
f taille de lime
e picadura de lima
i taglio di lima
n kap; vijlkap
d Feilenhieb

672 **cut** *v* **a mortise**
[car/chi]
f couper une mortaise
e cortar una mortaja
i tagliare una mortasa
n een gat steken
d ein Stemmloch ausstechen; ein Zapfenloch ausstechen

673 **cut-off angle grinder**
[gri/san]
f meuleuse-tronçonneuse angulaire

e troncadora angular
i troncatrice angolare
n haakse doorslijpmachine
d Winkeltrennschleifer

674 cut-off stand; stand for portable cut-off work
[cpt/gri/san]
f semelle d'appui pour tronçonnage; support de tronçonnage
e pedestal para troncadora
i sostegno per troncatrice
n doorslijpstandaard
d Schlitten für Trennschleifen; Trennständer

675 cut-off stand with adaptor
[gri/san]
f support de tronçonnage avec adaptation
e pedestal para troncadora con adaptador
i sostegno per troncatrice con adattamento
n doorslijpstandaard met aanzetstuk
d Pendeltrennständer

676 cut-off table with adaptor
[gri/san]
f table de tronçonnage avec adaptation
e mesa de troncado con adaptador
i piano di lavoro per troncatura con adattamento
n doorslijpbank met aanzetstuk
d Trenntisch mit Vorsatz

677 cut-off wheel
[cpt/gri/met/san]
f meule de tronçonnage
e muela troncadora
i mola troncatrice
n doorslijpschijf
d Trennscheibe

678 cutter; cutting bit; milling cutter
[fre]
f fraise
e fresa
i fresa
n frees
d Fräser

679 cutter arbor; cutter holder; milling arbor
[fre]
f arbre porte-fraise; broche porte-fraise; mandrin de fraisage
e mandril portafresa; portafresa
i mandrino portafresa
n freesdoorn; freeshouder
d Fräsdorn; Frässpindel

* **cutter arbor → 81**

680 cutter file; milled file
[far/fre]
f fraiseuse-lime; lime fraisée; lime fraiseuse
e fresa-lima; lima fresada; lima fresadora
i fresa-lima; lima fresata; lima fresatrice
n freesvijl; gefreesde vijl
d Fräsfeile; gefräste Feile

681 cutter head
[fre]
f tête de moulurage
e cabeza de corte; cabezal portafresas
i testa portafresa
n freeskop; meskopfrees
d Fräskopf

682 cutter head knife
[fre]
f fraise à profiler
e fresa de perfiles
i fresa per profilare
n profielfrees
d Profilfräser

* **cutter holder → 679**

683 cutting beads
[car/pla]
f rabotage de boudins
e acepillado de junquillos
i piallatura di listelli
n kraalprofielen schaven
d Hobeln von Rundstäben

* **cutting bit → 678**

684 cutting die; die
[met/scr]
f filière à tarauder
e hilera de roscar
i filiera a filettare
n snijijzer
d Schneideisen

685 cutting edge
[chi/gen]
f tranchant
e filo
i filo
n snede
d Schneide

686 cutting edge; scythe edge
[agr]
f tranchant
e filo
i filo della falce

n zeissnede
d Dengel; Sensenschneide

687 cutting edge designs
[car/fre/woo]
f profils de fraises
e perfiles de fresas
i profili di frese
n freesprofielen
d Fräserprofile

688 cutting flutes
[car/pla]
f rabotage de rainures semirondes
e acepillado de ranuras de media caña
i piallatura di scanalature semitonde
n halfronde groeven schaven
d Hobeln von halbrunden Nuten

689 cutting ga(u)ge
[mea]
f trusquin à couteau
e galga cortante; gramil de cuchilla
i graffietto tagliante
n snijkruishout
d Messerstreichholz

690 cutting grooves
[car/pla]
f rabotage de rainures
e acepillado de ranuras
i piallatura di scanalature
n groeven schaven
d Hobeln von Nuten

691 cutting housings
[car/pla]
f rabotage d'entailles
e acepillado de incisiones
i piallatura d'intaccature
n kepen schaven
d Hobeln von Einkerbungen

692 cutting machine; milling machine
[fre]
f fraiseuse
e fresadora
i fresatrice
n freesmachine
d Fräsmaschine

693 cutting ovalo mouldings
[car/pla]
f rabotage de quarts de rond
e acepillado de óvolos
i piallatura di ovoli
n kwartronde kantprofielen schaven
d Hobeln von Viertelstäben

694 cutting punch
[gas]
f découpoir
e punzón cortador
i punzone da taglio
n uitkapper
d Aushauer

695 cutting rebates
[car/pla]
f rabotage de rebords
e acepillado de rebajos
i piallatura d'incavi
n sponningen schaven
d Hobeln von Falzen

696 cutting reeds
[car/pla]
f rabotage d'une série de boudins
e acepillado de una serie de jonquillos
i piallatura di una serie di listelli
n tongprofielen schaven
d Hobeln von nebeneinander liegenden Rundstäben

697 cutting shears
[she]
f grésoir; pince-cisaille
e tijeras cortadoras
i cesoie di taglio
n afschraapschaar; afsnijschaar
d Abtrennschere

698 cutting table; milling table
[fre]
f table de fraisage
e mesa de fresadora
i tavola di fresatrice
n freestafel
d Frästisch

699 cyclomower
[agr/hor]
f faucheuse rotative
e segadora rotativa
i falciatrice giratoria
n cyclomaaier
d Rotormäher

700 cylinder bit
[dri]
f foret hélicoïdal; mèche hélicoïdale
e barrena espiral; broca salomónica
i punta elicoidale
n spiraalboor
d Spiralbohrer

701 cylinder grinding wheel
[gri/san]

f meule cylindrique
e muela cilíndrica
i mola cilindrica
n cilindrische slijpschijf
d zylindrische Schleifscheibe

702 **cylinder liner puller**
[pul]
f extracteur de fourrure de cylindre
e extractor de camisas de cilindro
i estrattore di camicie di cilindro
n bustrekker
d Zylinderbüchsenabzieher

* **cylinder spanner → 41**

703 **cylinder wrench; pipe wrench**
[hyd/spa]
f pince multiprise
e llave para tubos
i chiave per tubi
n pijptang
d Rohrzange

* **cylinder wrench → 41**

704 **cylindrical cutter; surface cutter**
[fre]
f fraise cylindrique
e fresa cilíndrica; fresa de planear
i fresa cilindrica
n cilinderfrees
d Planfräser; Stirnradfräser; Walzenfräser

705 **cylindrical grinding**
[gri]
f meulage cylindrique
e esmerilado cilíndrico
i smerigliamento cilindrico
n cilindrisch slijpen
d Zylindrischschleifen

D

706 **dapping block; doming block**
[gas]
f dé à emboutir
e dado de embutir
i tassello a fori tondi
n gatenank
d Vertiefstempel

707 **darby; darby float; patter**
[dec/mas]
f taloche
e fratás
i fratazzo
n raapbord
d Reibebrett

708 **Davy lamp; miner's lamp; safety lamp**
[min/saf]
f lampe de Davy; lampe de mineur
e lámpara de Davy; lámpara de minero
i lampada di Davy; lampada di minatore
n davylamp; mijnlamp
d Davysche Sicherheitslampe; Grubenlampe

709 **dead-smooth cut**
[far]
f taille superdouce; taille superfine
e picadura superior
i taglio finissimo
n fijnste kap
d Doppelschlichthieb; Feinschlichthieb; Hieb 4

710 **deburring by filing; dressing by filing**
[far/met]
f ébarbage par limage; ébavurage par limage
e desbarbado por limadura
i sbavatura per limatura
n afbramen door vijlen
d Entgratung durch Feilen

711 **deburring file; dressing file**
[far]
f lime à ébarbage; lime à ébavurage
e lima desbarbadora
i lima sbavatrice
n aanzetvijl
d Ansatzfeile

712 **deburring rasp; dressing rasp**
[far]
f râpe à ébarbage; râpe à ébavurage
e escofina desbarbadora
i raspa sbavatrice
n aanzetrasp
d Ansatzraspel

713 **deburring wheel**
[cpt/gri/met]
f meule d'ébarbage
e muela desbarbadora
i mola da sbavare
n afbraamschijf
d Schruppscheibe

714 **decorator's brush** *US*; **paperhanger's brush** *GB*
[dec]
f brosse à encoller; brosse de tapissier
e brocha de tapicero
i pennello di tapezziere
n behangersborstel
d Kleisterbürste; Tapezierbürste

715 **decorator's folding rule** *US*; **paperhanger's folding rule** *GB*
[dec]
f mètre-pliant de tapissier
e metro plegable de tapicero
i metro pieghevole di tapezziere
n behangersduimstok
d Tapezierrollstock

716 **decorator's scissors** *US*; **paperhanger's scissors** *GB*
[dec]
f ciseaux de tapissier
e tijeras de tapicero
i cesoie di tapezziere
n behangersschaar
d Tapezierschere

717 **dee handle**
[agr/gen/hor]
f poignée en D
e mango en D
i manico in D
n d-greep
d D-Griff

718 **deep hole drilling machine; gun drill**
[dri]
f perceuse à trous oblongs
e taladro de cajear; taladro para agujeros profundos
i trapano per trivellazione profonda
n kanonboor; langgatboormachine; loopboormachine
d Langlochbohrmaschine; Tieflochbohrmaschine

719 **dendrometer**
[foy/mea]

f dendromètre
e dendrómetro
i dendrometro
n boomdikte- en hoogtemeter; dendrometer
d Baummesser; Dendrometer

720 depressed-center(re) grinding wheel
[gri/san]
f meule à moyen déporté
e muela de centro avellanado
i mola a centro sprofondato
n slijpschijf met verzonken midden
d gekröpfte Schleifscheibe; Schleifscheibe mit gekröpfter Mitte

721 depth finder; sounder
[mea/nav]
f sondeur
e sondador; sondeador
i scandaglio; sondatore
n dieplood
d Tieflot

722 depth ga(u)ge; drilling depth ga(u)ge
[dri]
f butée de profondeur
e tope de profundidad
i arresto di profondità
n diepteaanslag
d Tiefenanschlag

723 depth ga(u)ge
[mea]
f calibre pour profondeur
e calibre para profundidad
i calibro per profondità
n dieptecaliber
d Tiefenlehre

724 depthing router; hand router; router plane
[pla]
f guimbarde
e guimbarda
i pialla di fondo; sponderuola
n grondschaaf
d Grundhobel

725 depth of sawing cut
[saw]
f profondeur de coupe
e profundidad del corte de sierra
i profondità del tratto di sega
n diepte van de zaagsnede
d Schnittiefe

726 diagonal cutting nippers; diagonal cutting pliers
[pli]
f pince coupante diagonale
e alicate(s) de corte diagonal
i pinze per taglio diagonale
n diagonaal snijdende tang; diagonaaltang
d Diagonalschneider

727 diagonally milled toothing
[far]
f denture à fraisage diagonal
e dentado diagonal; dentadura de fresado diagonal
i dentatura a fresatura diagonale
n diagonaal gefreesde tanding
d diagonal gefräste Verzahnung

728 diamond dressing file; industrial diamond dressing file
[far]
f lime diamantée
e lima diamantada
i lima a grani di diamante
n diamantvijl
d Diamantfeile

729 diamond drill
[dri/min]
f sonde à pointe diamantée
e sonda de punta diamantada
i sonda a punta di diamante
n diamantboor
d Diamantbohrer

* **diamond drilling bit → 617**

730 diamond point chisel
[chi/met/sto]
f grain d'orge
e cincel de punta rómbica; puntero
i scalpello a punta rombica
n facetbeitel
d Rautenstichel; Vierkantstichel

731 diamond point chisel; spear point chisel
[chi/wool]
f ciseau pointu
e cincel de punta de lanza
i scalpello a punta di lancia
n groefbeitel; puntbeitel; puntijzer
d Spitzbeitel

732 diamond saw
[saw]
f scie diamantée
e sierra de diamante
i sega a polvere di diamante
n diamantzaag
d Diamantsäge

733 diastimeter; distance meter
[mea/sur]
f diastémomètre; diastimomètre; diastomètre
e diastímetro
i distanziometro
n afstandsmeter; diastimeter
d Diastimeter; Distanzmesser

734 dibber; dibble; planting stick
[agr/hor]
f plantoir
e plantador
i foraterra
n pootstok
d Pflanzbohrer; Setzholz

735 die; drawing die
[cpt/dra/met]
f filière à étirer
e hilera de estirar
i filiera per trafilatura
n treksteen
d Ziehstein

736 die; threading die
[cpt/met/scr]
f filière à fileter
e hilera roscadora; terraja
i filiera tonda per filettare
n snijplaat
d Gewindeschneideisen

* **die → 684**

737 die set; subpress
[met]
f porte-matrice
e juego de matrices; portamatriz
i portamatrice
n matrijsdrager
d Matrizenhalter; Schabotteneinsatz

738 die stock
[cpt/met/scr]
f porte-filière
e portaterraja
i portafiliera
n snijplaathouder
d Schneideisenhalter

739 digging fork
[agr/hor]
f fourche à bêcher
e horquilla de cavar; laya
i forca per scavare
n spitriek
d Grabegabel

740 digital rule
[mea]
f appareil de mesure digital
e aparato de medida digital
i strumento di misura digitale
n digitaal meetinstrument; direct afleesbaar meetapparaat
d digitales Messgerät

741 dinging hammer; flatter
[for/ham]
f marteau à dresser
e martillo de enderezar
i martello da spianare
n vlakhamer
d Richthammer

742 dish grinding wheel
[gri/san]
f meule assiette
e muela de cubeta
i mola piatta
n schotelslijpschijf
d Tellerschleifscheibe

* **distance meter → 733**

743 distemper brush; whitewash brush
[dec]
f brosse à blanchir
e cepillo de revocar
i pennello per imbiancare
n witkwast
d Weissbürste

744 ditch cleaning knife
[agr/hor]
f couteau de fossé
e cuchillo de zanja
i coltello di fosso
n haalmes; slootmes; walmes
d Grabenmesser

745 dividers
[mea]
f compas à pointes
e compás de puntas
i compasso tipo a punta
n steekpasser
d Spitzzirkel

* **DIY-tools → 50**

* **dog-leg chisel → 224**

746 dog-nail; dog-spike
[iwa/tra]
f crampin de rail
e escarpia de carril

i rampino
n spoorspijker
d Schienennagel

* **do-it-yourself tools → 50**

747 domed head
[chi/cpt]
f tête en forme de dôme
e cabeza en forma de cúpula
i testa in forma di cupola
n paddestoelvormige kop
d kuppelförmiger Kopf

748 dome of silence; silence dome
[hou/iwa]
f patin de chaise silencieux
e clavo de deslizamiento
i chiodo di scorrimento
n glijnagel; glijspijker
d Möbelgleiter

* **doming block → 706**

749 door latch
[hou/iwa/saf]
f loquet
e picaporte
i arresto di porta; saliscendi
n deurklink
d Türklinke

750 double angle cutter; equal angle cutter
[car/fre/woo]
f fraise angulaire à deux côtés
e fresa de ángulo de dos labios
i fresa angolare bilaterale
n frees met dubbele snijkant
d doppelseitiger Winkelfräser

751 double-bar ga(u)ge; double bevel
[car/meal
f fausse équerre double
e falsa escuadra doble; sentanilla doble
i squadra doppia a lati mobili; squadra falsa doppia
n dubbele zwaaihaak; dubbele zwei; dubbele zweihaak
d Doppelschmiege

* **double cut → 649**

752 double-edge axe
[axe]
f hache à double tranchant
e ancha de doble corte
i ascia a doppio bordo; ascia a doppio filo
n bikbijl
d zweischneidiges Beil

753 double-edged saw blade
[cpt/saw]
f lame à double tranchant
e hoja de sierra de dos filos
i lama di sega a doppio tagliente
n zaagblad met tanden met twee snijkanten
d Zweischneider

754 double-faced hammer
[ham]
f marteau à deux têtes
e martillo de dos cabezas
i martello a due teste
n hamer met twee koppen; tweekopshamer
d Hammer mit zwei Köpfen; Zweikopfhammer

755 double flute bit; double flute cutter
[car/fre/woo]
f fraise droite à deux cannelures
e fresa de dos cortes
i fresa a due tagli
n tweezijdige frees
d doppelseitiger Fräser; Zweischneider

756 double-handed saw; two-men cross-cut saw
[foy/saw]
f scie passepartout
e sierra para árboles
i sega a due mani; sega a traversa
n afkortzaag; trekzaag; tweemanstrekzaag
d Ablängsäge; Zugsäge

757 double-handed scraper; scraper plane
[car/pla]
f racloir
e cepillo raspador
i raschietto
n schraapschaaf
d Flachschaber

* **double hoe → 572**

758 double open-ended spanner
[spa]
f clé à fourche à deux becs; clé à fourche double
e llave fija de dos bocas
i chiave fissa doppia
n steeksleutel met dubbele bek
d Doppel-Gabelschlüssel

759 double plane iron
[pla]
f contrefer double
e contrahierro doble
i controferro doppio

n dubbele keerbeitel
d Doppelhobeleisen

760 **double-threaded screw; two-start thread**
[iwa/scr]
f vis à double pas
e tornillo de doble rosca
i vite a due filetti
n schroef met twee windingen
d doppelgängige Schraube

761 **dovetail; dovetail joint**
[car/joi/woo]
f assemblage à queue d'aronde
e ensambladura a cola de milano
i incastro a coda di rondine
n zwaluwstaartverbinding
d Schwalbenschwanz; Zinken

762 **dovetail bit; dovetail cutter**
[car/fre/woo]
f fraise queue d'aronde
e fresa para colas de milano
i fresa a coda di rondine
n zwaluwstaartfrees
d Zinkenfräser

* **dovetail joint → 761**

763 **dovetail marker; dovetail template**
[mea]
f gabarit pour queue d'aronde
e plantilla para cola de milano
i sagoma per coda di rondine
n zwaluwstaartmal
d Markierschablone

764 **dovetail miter(re); miter(re) dovetail; secret dovetail joint**
[car/joi/woo]
f assemblage à queue perdue; queue d'aronde cachée
e cola de milano oculta
i coda di rondine invisibile
n verdekte zwaluwstaartverbinding
d verdeckter Schwalbenschwanz; verdeckter Zinken

765 **dovetail saw**
[saw]
f scie pour couper les queues d'aronde
e sierra de endentar
i sega per tagliare code di rondine
n toffelzaag
d Zinkensäge

* **dovetail template → 763**

766 **dowel; rag bolt; stone bolt**
[fas/iwa]
f boulon de scellement
e perno de empotrado
i grappa
n dookbout
d Steinschraube

767 **dowel bit; duck's bill bit; spoon bit**
[dri]
f mèche à hélice à pointe de centrage et deux traçoirs
e barrena de cuchara
i punta a cucchiaio
n houtspiraalboor met centerpunt en voorsnijders
d Holzspiralbohrer mit Zentrierspitze und Vorschneider; Löffelbohrer

768 **dowel guide; dowel(l)ing jig; dowel locator**
[joi]
f gabarit à goujons
e guía de espijar
i maschera per mecchie per caviglie
n deuvelhulpstuk; deuvelmal
d Dübelbohrlehre

769 **dowel saw**
[saw]
f scie à chevilles
e sierra de tacos
i sega a caviglie
n tapzaag
d Dübelsäge

770 **dowel screw; hand-rail screw**
[iwa/scr]
f goujon fileté à deux bouts
e espiga roscada en ambos extremas
i caviglia filettata a due estremità
n stokschroef
d Dübelschraube

771 **downfeed system**
[dri/gen]
f système de translation
e sistema de avance vertical hacia abajo
i meccanismo per avanzamento
n schuifinrichting
d Vorschubmechanismus

* **drag bit → 252**

772 **drainage tube**
[agr/hor]
f drain
e tubo de drenaje
i tubo di drenaggio

n draineerbuis
d Dränrohr

773 draining spade
[agr/hor]
f louchet
e pala de drenaje
i vanga di drenaggio
n draineerspade
d Dränierschaufel

774 draw bench
[dra/met]
f tréfiloir
e banco para estirar
i trafila
n trekbank
d Ziehbank

775 drawer-front dovetail; lap dovetail
[car/joi/woo]
f assemblage à queue perdue
e ensambladura de cola de milano cubierta
i incastro a coda di rondine coperto
n verdekte zwaluwstaartverbinding
d Verbindung mit verdeckten Stosstellen

* **drawer lock chisel → 278**

776 draw filing; smooth finishing
[far/met]
f finissage lisse; finition lisse
e acabado fino
i finitura liscia
n glad afwerken
d Feinbearbeitung

777 draw hoe
[agr/hor]
f houe; hoyau; rayonneur simple
e aporcador; azada de arrastre
i zappa finestrata
n hak
d Hacke; Ziehhacke

778 drawhook
[foy/gen/iwa]
f harpon
e garfio
i gancio di trazione
n harpoenhaak; trekhaak
d Einreisshaken; Greifhaken

* **drawing die → 735**

* **drawing knife → 607, 780**

779 draw knife; draw shave
[coo]
f plane de tonnelier
e plana de tonelero
i coltello a petto
n keepmes; kerfbeitel
d Kerbschnitzmesser

780 draw knife; drawing knife
[car/woo]
f plane
e cuchilla de suavizar
i coltello da intaglio
n trekmes
d Reifmesser; Ziehmesser; Zugmesser

781 draw knife; peeling knife
[cut/foy]
f plane pour écorchage blanc
e cuchilla de descortezar
i coltello a due manici; coltello a petto
n boommes; trekmes
d Schälmesser; Schnitzmesser; Ziehmesser; Zugmesser

782 draw screw
[fou]
f piton d'enlèvement
e hembrilla de desmoldeo
i occhiello per il sollevamento
n uitlichtschroefoog
d Tragöse

* **draw shave → 779**

783 dresser
[cpt/for]
f tas
e tas
i tassello
n tas
d Tasso

784 dresser
[hyd]
f rabattoir
e martillo de planear el plomo
i martello per spianare il piombo
n loodklopper
d Bleiplätthammer; Bleischlichthammer

785 dressing
[gri/pol]
f dressage; rectification
e rectificación
i rettifica
n bijslijpen
d Geradeschleifen

786 dressing apparatus
[pla]

f gauchisseuse
e aparato de planeado
i apparecchio di spianamento
n planeerapparaat
d Abrichtgerät

787 dressing axe
[cut/sto]
f rustique
e trinchante
i ascia da tagliapietre
n bikhamer
d Flächeneisen

* **dressing by filing → 710**

* **dressing file → 711**

788 dressing pick
[cut/sto]
f pioche pointue
e dolobre
i gravina di ritoccatura
n steenhouweel
d Spitzhacke

* **dressing rasp → 712**

789 drift hammer
[ham]
f masse
e mazo
i martello cacciacuneo
n drijfhamer
d Treibfäustel

790 drill; drill bit
[cpt/dri]
f foret; mèche
e barrena; broca
i punta da trapano
n boor
d Bohrer

791 drill; drilling machine
[dri]
f perceuse
e taladradora; taladro
i trapanatrice; trapano
n boormachine
d Bohrmaschine

792 drill bit
[cpt/dri/min]
f trépan
e sonda; trépano
i sonda
n steenboor
d Bohrkrone; Gesteinsbohrer

793 drill bit; grooved bit; twist bit
[cpt/dri]
f foret hélicoïdal; mèche hélicoïdale
e broca espiral; broca helicoidal; broca salomónica
i punta elicoidale
n spiraalboor
d Spiralbohrer

* **drill bit → 242, 790**

794 drill bit diamond type
[dri/min]
f perceuse à diamant
e taladro con diamante
i trapano con diamante
n boordiamant; diamantboor
d Diamantbohrer

795 drill bit ejector
[cpt/dri]
f éjecteur de trépans
e arrancasondas; sacabarrenas; sacatrépanos
i eiettore di sonde
n steenbooruitwerper
d Bohrerausheber; Bohrkronenausheber

796 drill bit extension
[min/pha]
f rallonge de trépans
e prolongación de sondas; prolongación de trépanos
i prolungamento di sonde
n steenboorverlenging
d Bohrerverlängerung; Bohrkronenverlängerung

797 drill bit roller type; roller bit
[dri/min]
f trépan à rouleaux en métal dur
e sonda de rodillos; trépano de rodillos
i sonda a rulli
n steenboor met hardmetaalrollen
d Drehbohrer; Gesteinsbohrer mit Hartmetallrollen

798 drill bits case
[cpt/dri]
f étui à forets
e caja de brocas
i astuccio per punte; scatola per punte
n boorcassette
d Bohrerschachtel

799 drill bit sharpener; drill-grinding jig
[cpt/dri/sha]
f machine à affûter les forets hélicoïdaux; machine à affûter les mèches hélicoïdales

e rectificadora de brocas espirales
i affilatrice di punte elicoidali
n spiraalborenslijpmachine
d Spiralbohrer-Schleifmaschine

800 drill bit with right-hand twist
[dri]
f foret à hélice à droite
e broca helicoidal cortanda a la derecha
i punta elicoidale a destra
n rechtssnijdende spiraalboor
d rechtsschneidender Spiralbohrer

801 drill bit with tungsten carbide negative cut tip
[dri]
f foret à pastille en carbure de tungstène à coupe négative
e broca con uña con metal extraduro de corte negativo
i punta a pastiglia al carburo di tungsteno per taglio negativo
n boor met hardmetaalsnijplaatje voor negatieve snede
d Bohrer mit Hartmetallplättchen für negativen Schnitt

* **drill borer → 306**

* **drill brace → 248**

802 drill bush
[cpt/dri]
f douille porte-foret
e casquillo-guía de portabroca
i bussola per foratura
n boorhuls
d Klemmhülse

803 drill chuck
[cpt/dri]
f porte-foret
e portabroca
i portapunta
n boorhouder
d Bohrerhalter; Spannfutter

804 drill chuck guard
[cpt/dri]
f protection de mandrin
e protector de portabroca
i protezione per portapunta
n boorhouderafscherming
d Bohrfuttersicherung

* **drill-grinding jig → 799**

805 drill head
[cpt/dri]
f tête de porte-foret
e cabezal de portabroca
i testa di portapunta
n boorkop
d Bohrfutterkopf

806 drill head key
[cpt/dri]
f clé de tête de porte-foret
e llave de cabezal de portabroca
i chiave di testa di portapunta
n boorkopsleutel
d Bohrfutterschlüssel

807 drill hole
[dri]
f trou de forage; trou de perçage
e pozo taladrado
i pozzo trapanato
n boorgat
d Bohrloch

808 drill hole cleaner
[cpt/dri]
f grattoir du trou de forage; grattoir du trou de perçage
e rasguño del pozo taladrado
i raschietto del pozzo trapanato
n boorgatkrabber; boorkrabber
d Bohrlochkratzer

809 drilling bit
[chi/dri/min]
f tête foreuse; trépan de sondage
e trépano de sondeo
i scalpello da perforazione
n beitelboor; boorbeitel
d Bohrmeissel

* **drilling depth ga(u)ge → 722**

810 drilling jig
[dri/meal]
f gabarit d'alésage
e calibre de taladros; galga para brocas
i calibro per fori; calibro per punte di trapano
n borenplaat
d Bohrungslehre

* **drilling machine → 791**

811 drilling machine standard
[cpt/dri]
f colonne de perceuse; montant de foreuse
e columna de taladro; montante de taladro
i montante di trapano
n boormachinestandaard
d Bohrmaschinenständer

812 **drilling rasp**
[dri/far]
f lime rifloir
e lima alisadora
i lima alesatrice
n schrobrasp
d Bohrsäge

813 **drill size ga(u)ge; number sizes**
[dri]
f calibre de perçage
e galga de agujeros
i calibro per punte di trapano
n gatkaliber
d Millimeter-Lochlehre

814 **drill spindle**
[cpt/dri]
f barre de forage; broche de forage
e árbol portabrocas; husillo de taladro
i albero di trapano; mandrino di trapano
n booras; boorspil
d Bohrspindel

815 **drill spoon**
[cpt/dri]
f cuiller de foret; culette
e cuchara de taladro
i cucchiaio di trapano
n boorlepel
d Bohrlöffel

816 **drill stand**
[dri]
f support de perçage
e bastidor de taladro
i supporto verticale per trapani
n boormachinestandaard
d Bohrständer

817 **drill travel**
[dri]
f course de perçage
e carrera de perforación
i corsa di perforazione
n boorverplaatsing
d Bohrhub

818 **driven-in wedge**
[cpt/foy]
f coin enfoncé
e cuña clavándose
i cuneo spinto in dentro
n ingedreven wig
d eingetriebener Keil

819 **drive screw; hammer-drive screw; screw nail**
[iwa]
f clou fileté
e clavo roscado
i chiodo a vite
n schroefnagel
d Schraubnagel

820 **drive spindle**
[cpt/gen/san]
f axe de raccordement
e husillo de mando
i albero di comando
n aandrijfas
d Anschlusswelle

821 **drop hammer**
[ham]
f mouton à chute libre
e martinete de caída libre
i maglio a caduta libera
n valhamer
d Fallhammer

822 **drunken saw; pendulum circular saw; wobbling saw**
[saw]
f scie circulaire oscillante
e sierra circular de paralelogramo
i sega circolare a pendolo
n slingerende cirkelzaag; slingerzaag
d Pendelkreissäge

823 **drunken thread; multiple thread**
[iwa/scr]
f filet à pas multiple
e rosca de pasos múltiples
i filettatura multipla
n meergangige winding
d vielgängiges Gewinde

824 **dual grip handle**
[cpt/dri]
f poignée à deux positions
e empuñadura de dos posiciones
i impugnatura a due posizioni
n handgreep met twee standen
d Handgriff mit zwei Positionen

825 **duckbill pliers; flat nosed pliers; flat nose pliers; pendulum pliers; square nosed pliers**
[pli]
f pince à courber à becs plats; pince à plier à becs plats
e alicate(s) para curvar de bocas planas; alicate(s) para doblar de bocas planas
i tenaglie di curvatura a becchi piatti
n buigtang met platte bekken
d Biegezange mit flachen Backen

* **duck's bill bit → 767**

826 dull grinding; matt(e) grinding
[gri/pol]
f dépolissage
e esmerilado sin brillo
i smerigliatura opaca
n matslijpen
d Mattschleifen

827 dung hoe
[agr/hor]
f croc à fumier
e horca para el estiércol
i forcone ricurvo per letame
n mesthak
d Misthacke

828 dust catcher
[cle/gen]
f pare-poussière
e recogedor de polvo
i raccoglitore di polvere
n gruisvanger
d Staubfänger

829 dust filter
[cle/gen]
f filtre à poussière
e filtro de polvo
i filtro da polvere
n stoffilter
d Staubfilter

830 dusting brush
[cle/hou]
f balai
e escobilla de mano
i spazzola
n stofbezem; stoffer
d Staubbürste

831 dusting brush; jamb brush
[cle/dec]
f brosse à poussière
e cepillo de polvo
i pennello per polvere
n stofkwast
d Staubbürste

832 dust mask
[saf]
f masque à poussière
e máscara contra polvo
i maschera di polvere
n stofmasker
d Staubmaske

833 dust pan
[cle/hou]
f pelle à poussière
e cogedor
i paletta per la spazzatura
n vuilnisblik
d Kehrschaufel

834 dust-proof cover
[gen/saf]
f capuchon pare-poussière
e casquete de protección contra polvo
i copertina a prova di polvere
n stofbeschermkap
d Staubschutzkappe

E

835 **ear muffs**
[saf]
f protège-oreilles
e orejeras
i paraorecchi
n oorbeschermers
d Ohrenschützer

836 **earth grab**
[civ/dre/min]
f benne piocheuse
e cuchara de excavadora
i cucchiaio di scavatrice
n graafemmer
d Greifbagger

837 **earth mover**
[civ/dre]
f excavatrice
e excavadora
i scavatrice
n graafmachine
d Bagger

838 **easel; studio easel**
[art]
f chevalet d'atelier
e caballete de estudio
i cavalletto da studio
n ezel; schildersezel
d Atelierstaffelei; Gestell

839 **easing wedges; lowering wedges; striking wedges**
[gen]
f coins d'ajustage
e cuñas para descrimbar
i cunei di aggiustaggio
n verstelwiggen
d Verstellkeile

840 **eccentric chuck; eccentric mandril**
[cpt/dri]
f mandrin excentrique
e mandril excéntrico
i mandrino eccentrico
n excentrische boorhouder
d exzentrisches Spannfutter; Versetzkopf

* **edge clamp → 621**

* **edge cramp → 621**

841 **edge joint**
[car/joi/wool]
f assemblage à chanfreins
e junta achaflanada; junta biselada
i unione a coltello di due assi
n schuine las
d Eckverband

842 **edge nailing**
[car/wool]
f clouage invisible
e clavado invisible
i inchiodatura nascosta
n onzichtbaar spijkeren
d verdecktes Nageln

843 **edge of a file**
[far]
f bord de lime; chant de lime
e canto de lima
i bordo di lima
n kant van een vijl
d Feilenseite

* **edge plane → 363**

844 **edger; edging trowel**
[mas]
f truelle pour arrondir les bords
e llana para redondear esquinas
i cazzuola bordatrice
n randplakspaan
d Abrundkelle

845 **edge shears; edging shears**
[agr/hor]
f cisailles à bordures
e tijeras de podar arriates
i forbici tosaerba per bordura
n graskantenschaar
d Rasenkantenschere

846 **edge shot**
[car/wool]
f planche à un bord raboté
e arista matada
i asse con un bordo piallato
n aan één kant geschaafde plank
d einseitig gehobeltes Brett

847 **edge tool**
[fre]
f fraise à trous oblongs
e fresa de cojear
i fresa per fori profonde
n frees voor diepe gaten
d Langlochfräser

848 **edge tool**
[cut]
f outil tranchant
e útil para cortar

i utensile da taglio
n snijgereedschap
d Schneidwerkzeug

849 **edge trimmer**
[agr/hor]
f coupe-bordures; hache dresse-bordures
e cortacepellones
i tagliabordi
n graskantensteker
d Rasenkantenstecher

850 **edge trimming cutter**
[fre]
f affleureuse
e fresa de achaflanar; fresa de cantear
i fresa di bordi
n kantfrees
d Kantenfräser

851 **edge trimming of joints**
[car/pla/wool]
f affleurage des assemblages
e recortado de juntas
i rifinitura di incastri
n bijschaven van verbindingen
d Bundighobeln von Verbindungen

852 **edge trimming plane**
[pla]
f rabot à polir les bords
e cepillo de cantear
i pialletto a rifinire bordi
n randschaaf
d Kantenhobel

853 **edging clamp**
[cla/iwa]
f serre-joint à coins
e sargento de bordes
i morsetto doppio ad angoli per cornici
n kantklem
d Kantenzwinge

854 **edging machine**
[pol]
f machine à border
e rebordeadora
i bordatrice
n pijpomfelsmachine
d Bördelmaschine

855 **edging saw**
[saw]
f scie à rogner
e sierra canteadora; sierra de cantear
i sega per bordatura
n randzaag
d Besäumsäge

* **edging shears → 845**

856 **edging tool**
[cpt/for]
f bordoir
e rebordeador
i utensile bordatore
n felsijzer
d Bördeleisen

* **edging trowel → 844**

* **egg cal(l)ipers → 305**

857 **egg sleeker; spoon tool**
[fou]
f lissoir à cuiller
e alisador de cuchara
i cucchiaio; cucchiarozzo
n polijstlepel
d Polierlöffel

858 **ejection of sawdust**
[saw]
f évacuation des sciures
e evacuación del serrín
i soffiatura della segatura
n spaanderuitwerping
d Späneauswurf

859 **elbow; knee**
[hyd/joi]
f bâton rompu; tuyau en coude
e codo; tubo acodado
i raccordo a gomito
n elleboog
d Knie; Knierohr

860 **elbow wrench**
[spa]
f clé mâle coudée
e llave acodada
i chiave a pipa
n soksleutel
d gebogener Steckschlüssel; Pfeifenkopfschlüssel

861 **electric cleaner; suction cleaner; vacuum cleaner**
[cle/hou]
f aspirateur de poussière
e electroaspirador de polvo; vacuoaspirador de polvo
i aspirapolvere
n stofzuiger
d Staubsauger

862 **electric cooker**
[hou/nut]

f cuisinière électrique
e cocina eléctrica
i cucina elettrica
n elektrisch fornuis
d elektrischer Küchenherd

863 electric crane
[tra]
f grue électrique
e electrogrúa
i gru elettrica
n elektrische kraan
d Elektrokran

864 electric drilling machine
[dri]
f perceuse électrique à main
e taladro eléctrico de mano
i trapano elettrico a mano
n elektrische handboormachine
d Elektrohandbohrmaschine

865 electric hoist
[tra]
f palan électrique
e elevador eléctrico; montacargas eléctrico
i paranco elettrico
n elektrisch hefwerktuig
d Elektrohebezeug; Elektrozug

866 electrician's chisel; floor board chisel
[chi]
f ciseau à soulever les planches du sol
e cincel arrancador
i paletta di ferro per staccare le assi del pavimento
n vloerbeitel
d Beitel zum Emporheben von Flurbrettern

867 electrician's pliers
[pli]
f pince d'électricien
e alicate(s) de electricista
i pinza di elettricista
n elektriciënstang; monteurstang
d Elektrikerzange

868 electric immersion heater
[hou/nut]
f thermoplongeur
e calentador eléctrico de inmersión
i scalda-acqua ad immersione
n dompelelement
d Tauchsieder

869 electric iron
[hou]
f fer à repasser électrique
e plancha eléctrica
i ferro da stiro elettrico
n elektrisch strijkijzer
d elektrisches Bügeleisen

870 electric soldering iron
[ele/wel]
f fer à souder électrique
e electrosoldador
i ferro da saldatura elettrico
n elektrische soldeerbout
d Elektrolötkolben

871 electrode carrier; electrode holder
[ele/wel]
f porte-électrode
e portaelectrodo
i portaelettrodo
n elektrodenhouder; lastang
d Elektrodenhalter

872 electromagnetic crane
[tra]
f grue électromagnétique
e grúa electromagnética
i gru elettromagnetica
n elektromagnetische kraan
d elektromagnetischer Kran

873 electromagnetic hammer
[ele/ham/lea]
f marteau électromagnétique
e martillo electromagnético
i martello elettromagnetico
n elektromagnetische hamer
d elektromagnetischer Hammer

874 electrometer
[ele/mea]
f électromètre
e electrómetro
i elettrometro
n elektrometer
d Elektrometer

875 electronic pliers
[pli]
f pince pour l'électronique
e alicate(s) para la electrónica
i pinza per la elettronica
n elektronikatang
d Elektronikzange

876 electropneumatic drill and chipping hammer
[ham/pha]
f marteau électropneumatique perfo-burineur
e martillo electroneumático perforador-burinador

i martello elettropneumatico perforatore a percussione
n elektropneumatische boor- en slaghamer
d elektropneumatischer Bohr- und Schlaghammer

877 electropneumatic hammer
[ham/pha]
f marteau électropneumatique
e martillo electroneumático
i martello elettropneumatico
n elektropneumatische hamer
d elektropneumatischer Hammer

878 elevating screw
[cpt/ele/scr]
f vis de pointage en hauteur
e tornillo de puntería en elevación
i vite di elevazione
n richtschroef
d Höhenstellspindel; Richtschraube

879 elevator
[tra]
f monte-charge
e elevador
i montacarichi
n hefwerktuig
d Elevator; Förderwerk; Hebewerk; Lastaufzug

880 elevator bucket
[dre]
f godet d'élévateur
e cangilón de elevador
i tazza per elevatore
n hijsemmer
d Hebewerkeimer

881 elevator chain
[cpt/dre]
f chaîne d'élévateur
e cadena de elevador
i catena di elevatore
n hefwerkketting
d Hebewerkkette

882 embossing hammer
[ham/met]
f marteau à emboutir; marteau de fond
e martillo de embutir en relievo
i martello battilastra
n bosseleerhamer
d Tiefhammer

883 emery buff
[pol]
f buffle émerisé
e cuero de búfalo esmerilado
i bufalo smerigliato
n met amarilpoeder bedekt buffelleer
d mit Schmirgel bedecktes Büffelleder

884 emery cloth
[pol]
f toile émerisée
e tela de esmeril
i tela smeriglio
n schuurlinnen
d Schmirgelleinen

885 emery paper
[gri/san]
f papier à l'émeri
e papel esmeril
i carta smeriglio
n amarilpapier
d Schmirgelpapier

886 emery paste
[che/pol]
f pâte d'émeri
e pasta esmeril
i spoltiglio
n amarilpasta
d Schmirgelpaste

887 emery powder
[che/pol]
f poudre d'émeri
e polvo de esmeril
i polvere di smeriglio
n amarilpoeder
d Schmirgelpulver

888 emery stone
[che/pol]
f pierre à émeri
e piedra de esmeril
i cote da smerigliare
n amarilsteen
d Schmirgelstein

889 emery wheel
[che/pol]
f meule d'ébauchage; meule-émeri
e muela de esmeril
i mola smeriglio
n amarilschijf; schuurschijf
d Polierscheibe; Schmirgelscheibe

*** enchasing hammer → 475**

890 end bench hook
[car/cpt]
f presse d'arrière d'établi
e tope final del banco
i cane finale del banco

n achterklauw; achtertang
d Hinterzange

891 **end block; end ga(u)ge**
[meal]
f cale-étalon
e bloquito calibrador; calibre de alturas; medida final
i blocco di taratura
n eindmaat
d Endmass; Parallelmass

892 **end cap**
[hyd]
f calotte obturatrice
e capacete; cofia
i tappo per tubi
n afsluitkap
d Abschlusskappe

893 **end cutter; end mill; end milling cutter**
[fre]
f fraise à queue; fraise en bout
e fresa con vástago; fresa de mango
i fresa a bottone; fresa codolo; fresa frontale
n kolffrees; vingerfrees
d Fingerfräser; Schaftfräser

894 **end cutting pliers; front cutting nippers; top cutting nippers**
[pli]
f pince coupante en bout
e alicate(s) de corte por delante; alicate(s) de corte frontal
i pinza a taglio frontale
n kopkniptang; voorsnijtang
d Vornschneider

* **end ga(u)ge → 891**

895 **end grinding wheel; lapidary grinding wheel**
[gri/san]
f meule lapidaire
e muela frontal; muela lapidaria
i mola frontale; mola lapidaria
n frontslijpschijf
d Stirnschleifscheibe

896 **end links**
[gen/iwa]
f anneaux terminaux; chaînons terminaux
e eslabones terminales
i anelli terminali
n eindschakels van een ketting
d Endglieder; Endschaken

* **end mill → 893**

* **end milling cutter → 893**

897 **end tenon**
[car/joi/woo]
f tenon en about
e muñón frontal
i tenone frontale
n koppen
d Stirnzapfen

898 **end vice**
[car/woo]
f boîte de rappel; presse de derrière; vis de derrière
e pinza trasera
i morsa a carrello; morsa posteriore
n achterklauw; achtertang
d Hinterzange

899 **engineer's chain; surveyor's chain**
[mea/sur]
f chaîne d'arpenteur
e cadena de agrimensor
i catena d'agrimensore
n meetketting
d Messkette

900 **engineer's file**
[far]
f lime d'ajusteur; lime d'atelier
e lima de ajustador; lima de mecánico
i lima di meccanico
n bankwerkersvijl
d Werkstattfeile

* **engineer's hammer → 206**

901 **engineer's pliers**
[pli]
f pince de mécanicien
e alicate(s) de mecánico
i pinza di meccanico
d Mechanikerzange

902 **engineer's sliding bevel**
[meal]
f fausse équerre de précision
e falsa escuadra de precisión
i squadra falsa di precisione
n precisiezwei
d Präzisionsschmiege

903 **engineer's try square**
[meal]
f équerre de précision
e gramil de precisión
i squadra a battente di precisione
n precisieschrijfhaak
d Präzisionswinkel

904 engraving ball
[gas]
f boule de graveur
e bola para grabar
i sfera per incidere
n graveerkogel
d Gravierkugel

905 entering tap; taper tap
[dri/scr]
f taraud ébaucheur; traçoir
e macho roscador cónico; primer macho de roscar
i maschio conico per filettare
n konische tap; voorsnijder
d Vorschneider

* **equal angle cutter → 750**

906 escapement of a plane; throat of a plane
[cpt/pla]
f ouverture de coin
e abertura de cepillo
i apertura di pialla
n bek van een schaaf
d Maul eines Hobels

907 etching needle
[art]
f burin
e buril
i bulino
n graveernaald
d Gravierstichel

908 even pitch
[iwa/lat/scr]
f pas constant
e paso constante
i passo costante
n constante spoed
d konstante Ganghöhe

909 expanding bit
[cpt/dri]
f foret à épaisseur réglable
e taladro de espesor regulable
i punta di spessore regolabile
n boor met verstelbare diameter
d Bohrer mit verstellbarem Durchmesser; Stellbohrer

910 expanding bottom reamer
[dri/fre/rea]
f alésoir expansible de fond
e escariador extensible de fondo
i alesatore ad espansione di fondo
n verstelbare bodemruimer
d nachstellbare Maschinengrundreibahle

911 expanding hand reamer
[dri/fre/rea]
f alésoir à main expansible
e escariador de mano de expansión; escariador de mano extensible
i alesatore a mano ad espansione
n verstelbare handruimer
d verstellbare Handreibahle

912 expanding machine reamer
[dri/fre/rea]
f alésoir équarrissoir; alésoir mécanique expansible
e escariador mecánico de expansión; escariador mecánico extensible
i alesatore meccanico ad espansione
n verstelbare machineruimer
d Maschinendehnreibahle; nachstellbare Maschinenreibahle

913 expanding mandril
[cpt/dri]
f mandrin à expansion
e mandril de expansión
i mandrino ad espansione
n verstelbare boorhouder
d ausdehnbarer Spindel; Spanndorn; Spreizdorn

914 expanding reamer with long knives
[dri/fre/rea]
f alésoir expansible à couteaux longs
e escariador extensible de cuchillas oblongas
i alesatore ad espansione con coltelli lunghi
n verstelbare ruimer met lange messen
d einstellbare Langmesser-Reibahle

915 expanding reamer with oscillating knives
[dri/fre/rea]
f alésoir expansible à couteaux oscillants
e escariador extensible de cuchillas oscilantes
i alesatore ad espansione con coltelli oscillanti
n verstelbare ruimer met excentrische messen
d verstellbare Reibahle mit Pendelmessern

916 expanding shell-shaped bottom reamer
[dri/fre/rea]
f alésoir expansible creux de fond
e escariador extensible hueco de fondo
i alesatore ad espansione cavo di fondo
n verstelbare opsteekbodemruimer
d nachstellbare Aufsteckgrundreibahle

917 expanding universal reamer
[dri/fre/rea]
f alésoir universel expansible
e escariador universal extensible
i alesatore universale ad espansione
n verstelbare universele ruimer
d einstellbare Universal-Reibahle

918 explosive rivet
[iwa/riv]
f rivet explosif
e remache explosivo
i ribattino esplosivo
n plofnagel
d Sprengniete

919 extension ladder
[fir/hou/saf]
f échelle coulissante
e escalera telescópica
i scala allungabile; scala estensibile
n schuifladder
d Ausziehleiter; Schiebeleiter

920 extension rule
[mea]
f jauge à coulisse
e regla corrediza
i riga ad estensione
n uitschuifbare duimstok
d verschiebbares Lineal

921 external cal(l)ipers; external ga(u)ge; outside cal(l)ipers
[mea]
f compas d'épaisseur; maître à danser
e calibrador exterior; compás testigo
i calibro per alberi; compasso di spessore
n dansmeester
d Greifzirkel; Rundtaster; Tanzmeister; Tastzirkel

922 external screw thread; external thread
[gen/scr]
f filetage extérieur
e rosca de tornillo macho
i filettatura esterna
n uitwendige schroefdraad
d Aussengewinde

923 eye bolt
[cpt/iwa]
f boulon à œillet
e macho de anillo
i bullone ad occhio; bullone a snodo
n oogbout
d Ösenbolzen; Ringbolzen

924 eye bolt with collar
[fas/iwa]
f boulon à œil à col
e perno de ojo con collar
i bullone ad occhio con collare
n oogbout met kraag
d Ösenbolzen mit Kragen

925 eye guards
[gen/saf]
f lunettes de protection
e gafas protectoras
i paraocchi
n oogbeschermers
d Schutzbrillen

926 eyelet pliers; eyelet punch
[pli]
f pince à poser les œillets
e ojeteador; sacabocado(s)
i pinza occhiellatrice
n oogjestang; ponstang
d Ösenzange

F

927 **face; working surface**
[cpt/gen]
f face de travail
e plano de trabajo
i piano di lavoro
n werkzaam vlak
d Wirkfläche

928 **face cutter**
[fre]
f fraise de face; fraise en bout
e fresa frontal
i fresa frontale
n kopfrees
d Stirnfräser

929 **face hammer**
[ham]
f marteau à panne plate
e martillo de cotillo plano; martillo de peña plana
i martello a penna piatta
n hamer met platte pen
d Hammer mit flacher Pinne

930 **face of a file**
[far]
f face de lime
e superficie de lima
i superficie di lima
n vlak van een vijl
d Feilenfläche

931 **face of a hammer; hammer face**
[ham]
f face du marteau; frappe du marteau
e boca del martillo; tabla del martillo
i bocca del martello
n hamerbaan; hamervlak
d Hammerbahn

932 **facing countersink**
[dri/rea]
f alésoir-fraise de dressage
e avellanador para refrentar
i accecatoio da spinare
n uitboorverzinkfrees
d Flachsenker

933 **facing cutter**
[fre]
f fraise à surfacer
e fresa de refrentar
i fresa da spianare
n afvlakfrees
d Flächenfräser

934 **fall bar**
[hou/saf]
f verrou tombant
e cerrojo de caída
i catenaccio da dente
n valgrendel
d Fallriegel

935 **falsework hammer; scaffolding hammer**
[bui/ham]
f marteau d'échafaudage
e martillo para andamios
i martello d'impalcatura
n steigerhamer
d Gerüsthammer

936 **fang bolt**
[iwa]
f boulon avec écrou à ergots; boulon d'ancrage
e perno arponado; perno para empotrar
i bullone di ancoraggio
n ankerbout
d Ankerbolzen

937 **fantail burner; streamline burner**
[bur]
f brûleur à charbon pulvérisé
e mechero de mariposa
i bruciatore a polvere di carbone
n waaierbrander
d Kohlenstaubbrenner

* **farmer's tools** → **33**

938 **farrier's hammer; shoeing hammer**
[for/ham]
f marteau de maréchal-ferrant
e martillo de herrador
i martello di maniscalco
n hoefhamer; hoefsmidshamer
d Hufhammer; Hufschmiedehammer

939 **farrier's knife**
[for]
f boutoir; rogne-pied
e cuchillo de herrero
i incastro di maniscalco
n hoefkapmes; hoefmes
d Hufschmiedemesser

940 **farrier's tongs**
[for]
f tricoises
e tenazas de herrador
i tenaglie di maniscalco
n hoefsmidstang; hoeftang
d Hufeisenzange

941 farrier's tool box
[for]
f boîte à outils du maréchal ferrant
e caja de herramientas del herrero
i cassetta attrezzi da maniscalco
n beslagbak
d Werkzeugkasten eines Hufschmieds

942 fascine knife
[foy]
f couteau à fasciner
e cuchillo de fajina
i coltello da fascina
n fascinemes; kapmes
d Faschinenmesser

943 fast and idle pulley
[cpt/tra]
f poulie fixe et folle
e polea fija y loca
i gola fissa e a folle; puleggia fissa e a folle
n vaste en losse katrol
d Fest- und Losrolle

* **fastening bolt → 595**

944 fastening nail; pilot nail; sprigging
[car/eng/woo]
f clou d'attache
e clavo para asegurar
i chiodo di attacco
n hechtspijker; sloopspijker
d Haftnagel

945 fast pulley
[cpt/tra]
f poulie fixe
e polea fija
i puleggia fissa
n vaste katrol
d Festrolle

* **faucet** *US* **→ 364**

* **faucet hole** *US* **→ 365**

* **faucet hole borer** *US* **→ 366**

946 fawn foot
[axe/cpt/foy]
f épaississement de manche
e espesamiento de mango
i spessamento di manico
n steelverdikking
d Stielverdickung

947 faying face; faying surface
[car/joi/woo]
f surface d'affleurage
e superficie de contacto; superficie de encoramento
i superficie di contatto
n verbindingsvlak
d Verbindungsfläche

948 feather
[cpt/sto]
f coin à fendre
e cuña de partir
i cuneo per fendere
n splijtwig
d Spaltkeil

949 feather
[car/joi/woo]
f languette
e lengüeta
i linguetta
n messing; veer
d dünner Spund; Feder

950 feather duster
[cle/hou]
f plumeau
e plumero
i piumino
n plumeau; stofkwast
d Staubbesen; Staubwedel

951 feather-edged board
[car/joi/woo]
f planche taillée en biseau
e tabla achaflanada
i asse a linguetta
n plank met messingrand
d Brett mit Federrand

952 feather joint
[car/joi/woo]
f assemblage à languette
e empalme de lengüeta postiza; ensambladura de falsa espiga
i unione a linguetta
n losse veer- en groefverbinding
d Federverbindung

953 feather tongue
[car/joi/woo]
f languette
e espiga; lengüeta
i linguetta
n veer
d Feder

954 feeding rod
[fou]
f barre de pompage
e barra de picado

i barra di pompatura
n pompstaaf
d Pumpstab

955 feeler ga(u)ge
[meal]
f calibre à la lame; espion; jauge d'épaisseur; pige
e calibre de espesores; galga de rendijas; lamina calibradora
i calibro a lama; calibro a spessori
n gleufmaat; meettong; voelmaat
d Dickenschablone; Fühlerlehre; Spion

* **feeler ga(u)ge → 379**

956 felling axe
[axe/foy]
f hache d'abattage
e hacha de tala
i ascia per abbattimento
n boombijl; houthakkersbijl; velbijl
d Fällaxt

957 felling axe handle
[axe/cpt/foy]
f manche de la hache d'abattage
e mango de la hacha de tala
i manico dell'ascia per abbattimento
n boombijlsteel; velbijlsteel
d Fällaxthelm; Fällaxtstiel

958 felling wedge
[foy]
f coin à abattre
e cuña de tala
i cuneo da abbattamento
n velwig
d Fällkeil

959 felt-fibre-backed disk
[cpt/gri/pol]
f disque feutre sur fibre
e disco de fieltro ad empaquetadura de fibra
i piattorello fibra
n viltschijf met vezelonderlaag
d Filzscheibe auf Fiber

960 female ga(u)ge; ring ga(u)ge
[meal]
f calibre annulaire
e calibre anular; calibre hembra
i calibro ad anello
n ringkaliber
d Ringkaliber

961 female screwthread; internal screwthread
[scr]
f filetage intérieur
e hilo de rosca interno
i filettatura interna
n inwendige schroefdraad
d Innengewinde

962 fence
[cpt/gri/too]
f centreguide
e contraguía
i contraguida
n tegengeleider
d Gegenleiter

963 fence
[gen/saf]
f butée
e tope
i arresto
n aanslag
d Anschlag

964 fencing pliers
[agr/foy/hor/pli]
f pince-marteau
e tenaza-martillo
i pinza-martello
n hamertang
d Hammerzange

965 ferrule
[cpt/iwa]
f virole
e virola
i ghiera; virola
n ring
d Ring; Zwinge

966 ferrule of a file
[far]
f virole de lime
e virola de lima
i ghiera di lima
n ring van een vijl; versterkingsring
d Heftzwinge

967 fertilizer broadcaster *GB*; **synthetic fertilizer broadcaster**
[agr/hor]
f distributeur d'engrais; centrifuge à trémie vibrante
e distribuidor de abonos artificiales
i distributore centrifugo di fertilizzanti
n kunstmeststrooier
d Kreiseldüngerstreuer

968 fibre sanding disk
[cpt/san]

f disque fibre à poncer
e disco fibra para lijar
i disco fibra per levigare
n schuurschijf met hard fiber rug; schuurschijf met vezelbewapening
d Fiberscheibe; Fiberschleifscheibe

969 fiddle drill
[dri]
f arçon; drille à archet et corde
e taladro de arco y cuerda
i trapano ad arco e corda
n drilboor; fidelboor; fiedelboor
d Bogenbohrer

970 fieldcrop sprayer
[agr/hor]
f pulvérisateur
e pulverizador
i polverizzatore
n veldspuit
d Feldspritze

971 file
[far]
f lime
e lima
i lima
n vijl
d Feile

972 file brush; file card
[cpt/far]
f brosse à limes; carde
e carda para limas; cepillo de limas
i pennello per lime; spazzola per lime
n vijlborstel; vijlenborstel
d Feilenbürste

*** file cut → 671**

973 file cutter
[far]
f tailleur de limes
e picador de limas
i tagliatore di lime
n vijlenkapper
d Feilenhauer

974 file cutting
[far]
f taillage de lime
e picadura de lima
i taglio di lima
n kappen van de vijl
d Feilbehauung

975 file dust
[far]
f limaille; limature
e limaduras; limalla
i limature
n vijlsel
d Feilicht; Feilspäne; Feilstaub

976 file guide
[far/saw]
f guide de lime
e guía de lima
i guida di lima
n vijlgeleider
d Feilenführer

977 file *v* off
[far]
f enlever à la lime; ôter à la lime
e desfilar; quitar con la lima
i levare colla lima
n afvijlen
d abfeilen

978 file *v* over
[far]
f repasser à la lime
e repasar con la lima
i ripassare colla lima
n bijvijlen; navijlen
d nachfeilen; überfeilen

979 file rasp
[far/far]
f lime mordante; lime-râpe
e lima-escofina
i lima-raspa
n raspelvijl; raspvijl
d Feilraspel; Raspelfeile

980 file-shape saw
[saw]
f scie en forme de lime
e sierra en forma de lima
i sega in forma di lima
n boogvijl
d feilenförmige Säge

981 file stroke
[far]
f coup de lime; trait de lime
e corte de lima
i colpa di lima; corsa di lima
n vijlstreek
d Feilstrich

982 file with lozenge-shaped cross-section
[far]
f lime à section losangique; lime pignon
e lima de sección losanjeada
i lima a sezione losanga

n vijl met ruitvormige doorsnede
d Rhombussägefeile; Feile mit rautenförmigem Querschnitt

983 file with oval cross-section; oval file
[far]
f lime feuille de sauge
e lima almendra; lima hoja de savía; lima ovalada
i lima ovale
n ovale vijl; vijl met ovale doorsnede
d Vogelzunge; Vogelzungenfeile

984 filing
[far]
f limage
e limadura
i limatura
n vijlen
d Feilen

985 filing block
[far/met]
f bloc à limer
e bloque para limar
i blocchetto per la limatura
n vijlblok
d Feilfutter; Feilholz

986 filing machine
[far]
f limatrice; limeuse; machine à limer
e limadora
i limatrice; macchina a limare
n vijlmachine
d Feilmaschine

987 filing pin; goldsmith's file
[far/gas]
f lime d'orfèvre
e lima de orfebrero
i lima di orafo
n goudsmidsvijl
d Goldschmiedefeile

988 filler rod; welding rod
[cpt/wel]
f baguette de soudage
e varilla de metal de aportación; varilla para soldar
i bacchetta di apporto; bacchante saldante
n lasstaaf
d Schweisstab

989 fillet
[pri]
f filet
e filete
i filetto
n filetijzer; pletstempel
d Filete

990 filling knife
[dec]
f couteau de peintre
e cuchillo de pintor
i coltello per pittore
n stopmes
d Malermesser

991 fillister; fillister plane
[pla]
f guillaume à semelle réglable
e guillame de base regulable
i caprugginatoio; sponderuola a canto
n boordschaaf
d Simshobel mit doppeltem Eisen und verstellbarer Sohle; Tiefbohrhobel

992 fillister-head screw
[iwa/scr]
f vis à tête ronde
e tornillo de cabeza redonda
i vite a testa bombata
n rondkopschroef
d Rundkopfschraube

* **fillister plane → 991**

993 findings ***US*****; slaughterman's tools** ***GB***
[nut]
f outils de l'abatteur
e herramientas del matarife
i arnesi del macellaio
n slachtgereedschappen
d Schlächterwerkzeuge; Schlachtwerkzeuge

994 fine boring
[dri]
f perçage de précision
e taladrado de precisión
i foratura di precisione
n precisieboren
d Feinbohren

995 fine finish machining; fine machining
[fin]
f finissage mécanique
e afinado mecánico
i rifinitura di precisione
n machinaal fijnen
d maschinelle Feinbearbeitung

996 fine hackle
[tex]
f affinoir; peigne fin; sérin fin
e rastrillo fino
i pettine in fino; pettine stretto

n fijne hekel
d Ausmachhechel; Feinhechel

* **fine machining → 995**

997 fine toothing
[cpt/saw]
f denture fine
e dentadura fina
i dentatura fina
n fijne vertanding
d Feinverzahnung

998 finger stop
[cpt/tool]
f butée ajustable
e tope ajustable
i arresto aggiustabile
n verstelbare aanslag
d verstellbarer Anschlag

999 finishing cutter; finishing milling cutter
[fre]
f fraise à finissage
e fresa para acabar
i fresa a finire
n fijnfrees
d Fertigfräser

1000 finishing file
[far]
f lime à finissage; lime à finition
e lima de acabado; lima de pulir
i lima finitrice
n afwerkvijl
d Putzfeile

1001 finishing grinding wheel
[cpt/gri/pol]
f meule de finition
e muela de acabado
i mola finitrice
n fijnslijpsteen
d Fertigschleifscheibe

* **finishing milling cutter → 999**

1002 finishing reamer
[dri/fin/fre/rea]
f alésoir finisseur
e escariador acabador; escariador de terminación; escariador repasador
i alesatore finitore
n fijnruimer
d Fertigreibahle; Nachreibahle

* **finishing tap → 297**

1003 finishing tool
[fin/pol]
f acier à finir
e herramienta acabadora
i utensile per rifinire
n fineerstaal
d Schlichtstahl

1004 finishing trowel; laying-on trowel; metal float; plasterer's trowel
[mas]
f plâtroire à poignée ouverte; plâtroire rectangulaire
e paleta rectangular
i cazzuola rettangolare
n plakspaan; pleisterspaan; plekspaan
d Glättekelle

1005 fire bucket
[fir]
f seau à incendie
e cubo de bombero
i secchia di pompiere
n brandemmer
d Feuereimer

1006 fire engine
[fir]
f pompe à incendie
e bomba de incendio
i pompa da incendio
n brandspuit
d Feuerspritze

1007 fire escape
[fir/saf]
f échelle de secours
e escalera de emergencia
i scala di sicurezza antiincendio
n brandladder
d Feuerleiter; Rettungsleiter

1008 fire extinguisher
[fir]
f extincteur d'incendie à main
e extintor de mano
i estintore a mano
n brandblusapparaat
d Feuerlöschgerät; Handfeuerlöscher

1009 firefighter's axe; fireman's axe
[axe/fir]
f hachette de pompier
e hacha de bombero
i piccozza da pompiere
n brandbijl; brandweermanbijl
d Feuerwehrbeil

1010 fire-hook
[cpt/fir]
f crochet à incendie
e gancho para derribar cosas incendiadas
i gancio d'incendio
n brandhaak
d Feuerhaken

1011 fire-hook; fire-rake
[hou]
f attisoir; croc à feu; pique-feu; râble
e atizador
i attizzatore
n helhaak; hellehaak; vuurhaak
d Feuerhaken

1012 fire-hose
[fir]
f boyau d'incendie; lance
e manga de riega
i manichetta
n brandslang
d Feuerwehrschlange

* **fireman's axe → 1009**

* **fire-rake → 1011**

1013 firing tools
[hou]
f outils de chauffage
e herramientas de calefacción
i utensili di riscaldimento
n stookgereedschappen
d Heizungsgeräte

1014 firmer chisel
[chi]
f fermoir
e formón
i scalpello da digrossare
n fermoorbeitel; hakbeitel; vermoorbeitel
d Stemmeisen

1015 firmer gouch
[chi]
f fermoir en forme de gouge
e gubia de boca arqueda
i sgorbia per digrossare
n fermoorguts; hakguts; vermoorguts
d Hohlstemmeisen

1016 fir-tree lifting drill
[dri/foy]
f foret élévateur de sapins
e broca elevadora de pinabetes
i trapano elevatore d'abeti
n denneboor
d Tannenbohrer

* **fishtail bit → 252**

1017 fitch
[dec]
f brosse en putois
e brocha de pelos de turón
i pennello a capelli di puzzola
n bunzinghaarkwast
d Iltishaarbürste

1018 fitter's hammer
[ham]
f marteau d'ajusteur; marteau de monteur
e martillo de ajustador; martillo de mecánico
i martello di aggiustatore; martello di montatore
n bankwerkershamer; installateurshamer
d Installateurshammer

1019 fitter's pincers; fitter's pliers; fitter's tongs
[hyd]
f tenaille(s) d'appareilleur; tenaille(s) de gazier; tenaille(s) de plombier-installateur
e tenaza de gasista
i tenaglie d'idraulico; tenaglie d'installatore
n fitterstang
d Installateurszange

1020 fixed eccentric
[cpt/eng]
f excentrique fixe
e excéntrico fijo
i eccentrico fisso
n vaste excentriek
d fester Exzenter

1021 fixed face pin spanner
[spa]
f clé à ergots
e llave de gancho con espigas
i chiave a pioli
n vaste pensleutel
d Gelenkstirnlochschlüssel

1022 fixed hook spanner; hook spanner
[spa]
f clé à ergot plat
e llave de gancho
i chiave a settore
n haaksleutel; vaste haaksleutel
d Hakenschlüssel

1023 fixed pulley
[cpt/tra]
f poulie fixe
e polea fija
i puleggia fissa

n vaste katrol
d feste Rolle

1024 fixings
[cpt/fas/joi]
f moyens de fixation
e medios de sujeción
i elementi di collegamento
n bevestigingsmiddelen
d Befestigungsmittel

1025 flail
[agr]
f fléau
e mayal
i coreggiato; trebbia
n dorsvlegel; vlegel
d Dreschflegel

1026 flail handle; flail helve; flail staff
[agr]
f manche de fléau
e mango de mayal
i manfano
n vlegelstok
d Flegelstiel

1027 flanged fixing nuts; flanged nuts
[cpt/iwa]
f flasques-écrous de serrage
e tuercas de reborde de fijación; tuercas tapón de fijación
i dadi a colletto di collegamento
n spanmoeren
d Spannflanschmuttern

1028 flange set square
[mea]
f équerre à bride
e escuadra de brida; escuadra para colocar bridas
i squadra a brida
n flenswinkelhaak
d Flanschwinkel

1029 flanging apparatus
[met]
f appareil à brider
e aparato de rebordear
i apparecchio da bordatura
n flensapparaat; ompersgereedschap
d Bördelgerät; Bördelwerkzeug

1030 flask; mo(u)lding box
[fou]
f châssis
e caja de moldear
i staffa
n vormkast
d Formkasten

1031 flat adze
[coo]
f erminette plate
e azuela plana
i ascia a filo diritto
n platte dissel
d Flachdexel

1032 flat beck iron
[cpt/for]
f tas plat; tasseau plat
e tas plano
i tassello piatto
n vlakke tas
d flacher Tasso

1033 flat brush
[dec]
f pinceau plat
e brocha plana; pincel plano
i pennello piatto
n platte kwast
d Flachpinsel

1034 flat chisel
[chi/gen/pha]
f ciseau plat
e cincel plano
i scalpello piatto
n platte beitel
d Flachmeissel

1035 flat cut-off grinding disk
[cpt/gri]
f plate de tronçonnage
e disco plano para dividir
i mola troncatrice piatta
n doorslijpschijf vlakke uitvoering; vlakke doorslijpschijf
d flache Trennscheibe

1036 flat-faced stake
[cpt/for]
f bigorne plate
e bigornia plana
i bicornia piatta
n vlakstaak
d flache Faust

1037 flat file
[far]
f lime plate
e lima plana
i lima piatta
n platte vijl
d Flachfeile

1038 flat gouge
[car/chi]
f gouge à cambrure faible; gouge plate
e gubia de combeo pequeño; gubia plana
i sgorbia ad incurvimento piccolo; sgorbia piatta
n guts met geringe welving; platte guts
d Flachhohlbeitel; Hohlbeitel mit geringer Wölbung

1039 flat grinding wheel; straight grinding wheel
[cpt/gri/san]
f meule plate
e muela plana
i mola piatta
n vlakke slijpschijf
d Geradeschleifscheibe

1040 flat head rivet
[iwa/riv]
f rivet à tête aplatie
e remache de cabeza aplastada
i ribattino a testa piatta
n klinknagel met platte kop
d Flachkopfniete

1041 flat heavy-duty vice; heavy-duty machine vice
[cpt/vic]
f étau plat renforcé
e mordaza de máquina para trabajos fuertes
i morsa parallela con rinforzi
n zware machinebankschroef
d verstärkter Maschinenschraubstock

1042 flat metallic brush
[cpt/gen/gri]
f brosse métallique plate
e brocha metálica plana
i spazzola metallica piatta
n platte staaldraadborstel
d flache Stahldrahtbürste

* **flat nosed pliers → 825**

* **flat nose pliers → 825**

1043 flat-oval brush
[cpt/gen/gri]
f pinceau plat oval
e pincel óvalo-plano
i pennello ovale-piatto
n plat-ovale kwast
d ovaler Flachpinsel

1044 flat pliers
[pli]
f pince plate
e alicate(s) cortaalambres; alicate(s) de boca plana
i pinza a punte piatte; pinza piatta; pinza tagliafili
n draadtang
d Drahtzange; Flachzange

1045 flat rasp
[far]
f râpe plate
e escofina plana
i raspa piatta
n platte rasp
d Raspe (flachstumpfe Form)

* **flat square → 417**

1046 flattening hammer
[ham/met]
f aplatissoir; aplatissoire
e martillo aplanador
i martello di appiattamento
n boothamer; botshamer
d Flachhammer; Streckhammer

1047 flatter
[for]
f chasse à parer
e aplanador
i martello di spianare
n vlakhamer
d Schlichthammer

1048 flatter
[met/wat]
f filière à ruban
e hilera para alambre plano
i trafila per nastri
n treksteen voor platte draad
d Flachdrahtzieheisen

* **flatter → 741**

1049 flat vice; machine vice
[cpt/vic]
f étau plat
e mordaza de máquina; tornillo portapieza
i morsa parallela
n machinebankschroef
d Maschinenschraubstock

1050 fleam tooth
[saw]
f dent de scie en forme de triangle isocèle
e diente de sierra en forme de triángulo isósceles
i dente di sega in forma di triangolo isoscele

n zaagtand in de vorm van een gelijkbenige driehoek
d Sägezahn in Form eines gleichschenkligen Dreiecks

1051 fleam toothing; straight toothing; peg toothing
[saw]
f denture droite
e dentadura en forma de triángulo isósceles; dentadura recta
i dentatura retta
n rechte vertanding
d Dreieckzahnung

1052 fleshing knife
[lea]
f écharnoir
e cuchillo descarnador
i scarnatoio
n slichtmes
d Ausfleischmesser

1053 flexible-body file; flexible file
[far]
f lime flexible; lime incassable
e lima flexible
i lima flessibile
n buigzame vijl
d biegsame Feile

1054 flexible drive; flexible shaft
[cpt]
f arbre flexible
e árbol flexible
i albero flessibile
n flexibele as
d Biegewelle; biegsame Welle

*** flexible file → 1053**

1055 flexible file blade frame
[cpt/far]
f monture cintrable pour plaques fraiseuses
e soporte ajustable para lima-fresa
i supporto aggiustabile per lima-fresa
n verstelbare freesvijlbladhouder
d verstellbarer Fräserfeilenblatthalter

1056 flexible filling knife
[dec]
f couteau de peintre à lame flexible
e cuchillo de pintor flexible
i coltello per pittore flessibile
n stopmes met buigzaam blad
d flexibles Malermesser

1057 flexible roller bearing
[cpt/eng]
f palier à billes flexible
e cojinete de bolillas flexible
i cuscinetto a sfere flessibile
n flexibel kogellager
d flexibles Kugellager

1058 flexible sander attachment
[san]
f adaptation ponceuse souple
e adaptador de lijadora flexible
i adattamento per levigatrice flessibile
n aanzetstuk voor flexibele schuurmachine
d elastischer Tellerschleifvorsatz

1059 flexible saw; folding saw; helicoidal saw
[saw]
f scie à fil hélicoïdal
e sierra de hilo helicoidal
i sega a filo elicoidale
n draadzaag
d Drahtsäge

1060 flexible screwdriver
[scd]
f tournevis flexible
e destornillador flexible
i cacciavite flessibile
n flexibele schroevedraaier
d flexibler Schraubendreher; flexibler Schrauber

*** flexible shaft → 1054**

1061 flexible ventilated sanding backing pad
[cpt/gri]
f plateau souple ventilé de ponçage
e plato flexible ventilado de lijadura
i piattorello flessibile ventilato per levigatura
n buigzame slijpschijfrug met ventilatiegroeven
d elastischer Schleifteller mit Belüftungsrillen

1062 flint paper; glass paper
[gri/san]
f papier de verre
e papel de silex pulverizado
i carta a selce polverizzata
n flintpapier
d Flintpapier

1063 float
[mas]
f taloche
e llana; trulla
i piattello d'intonaco

n pleisterspaan
d Glättekelle

1064 float-cut file
[far]
f lime à taille simple
e lima de picadura simple
i lima a taglio semplice
n vijl met enkelvoudige kap
d einhiebige Feile

1065 flogging chisel
[chi]
f ciseau à froid lourd
e cortafrío de rebabar
i scalpello freddo da stozzatura
n zware koudbeitel
d schwerer Kaltmeissel

1066 flogging hammer
[ham]
f marteau batteur
e martillo de fundidor
i martello da stozzatura
n hamer voor ruwe bewerking
d Rammhammer

* **floor board chisel → 866**

1067 floor cramp; flooring cramp
[car]
f serre-joint de plancher
e mordaza de piso
i morsetto di pavimento
n vloeraandrijver
d Fussbodenklemme

1068 flooring saw; slitting saw
[saw]
f scie à rainurer
e sierra para entarimar
i sega per fessurare
n sleufzaag
d Gratsäge; Nutsäge

1069 flooring saw blade
[cpt/saw]
f lame de scie à dents trempés
e hoja de sierra de dientes templados
i lama di sega con denti temperati
n zaagblad met geharde tanden
d Sägeblatt mit gehärteten Zähnen

* **flooring trowel → 587**

1070 flue brush
[cle/hou]
f hérisson
e escoba de deshollinar
i spazzola per canna fumaria
n rookkanaalbezem
d Stossbesen

1071 flush bolt
[iwa]
f boulon à tête encastrée
e perno de cabeza embutida
i bullone a testa incassata; bullone a testa rasa
n bout met ingelaten kop
d Bolzen mit Versenkkopf

1072 flute; fluting
[gen]
f cannelure; rainurage
e acanaladura; ranura
i scanalatura
n groef; van groeven voorzien
d Kannelierung; Nuten; Riffelung

1073 fluted dressing brick; fluted rubbing brick
[gri/pol]
f brique-dresseur rainurée
e piedra rectificadora ranurada
i pietra equilibratrice scanalata
n gegroefde schuursteen
d ausgekehlter Abrichtstein

* **fluting → 1072**

1074 fluting plane
[pla]
f guillaume à canneler
e cepillo bocel; guillame de acanalar
i pialla per scanalare
n canneleerschaaf
d Kannelierhobel

1075 fly cutter
[fre]
f fraise à fendre
e fresa hendidora
i fresa di fessura
n sleuffrees
d Schlitzfräser

* **flynut → 378**

1076 fodder cutter; fodder knife
[agr]
f broyeur à fourrage; coupe-fourrage
e cuchillo del forraje; pisador del forraje
i tagliacotica
n voersnijder
d Futtermesser

1077 fodder-cutting foot-iron
[agr]
f couteau à pied
e hierro de pie cortador de forraje
i tagliacotica a piede
n voetmes
d Fretmesser

1078 fodder-cutting roller
[agr]
f rouleau cisailleur
e cortador rotante del forraje
i falciatrice
n rolhak
d Rollenhacke

* **fodder knife → 1076**

1079 folding anvil stake
[for]
f enclumette à plier; outil à agrafer
e útil de engatillado; yunque portátil para plegar
i incudine portatile per piegare
n felsijzer; felsstaak
d Falzmeissel; Falzstock

1080 folding measure; folding rule; foot rule
[mea]
f mètre-pliant
e metro plegable
i metro pieghevole
n duimstok
d Gliedermassstab; Zollstock

* **folding saw → 1059**

1081 folding wedge; locking key
[car/cpt]
f angrois; clavette
e cuña de apriete
i cuneo di fissaggio
n opsluitwig
d Schlusskeil; Vorsteckkeil

1082 foot blower; treadle blower
[cpt/for]
f soufflerie à pédale
e soplete de pedal
i mantice a pedale
n met de voet bediende blaasbalg
d Tretgebläse

1083 footprint wrench; pipe wrench
[spa]
f clé à tubes
e llave para tubos
i chiave serratubi
n gaspijpsleutel
d Rohrschlüssel

* **foot rule → 1080**

1084 fore plane
[car/pla/woo]
f demivarlope
e galera media
i pialla da sgrossare media; sgrossino medio
n kleine rijschaaf
d Schropphobel

1085 forge; smithy
[for]
f forge
e forja; fragua; taller de forja
i fucina
n smederij; smidse
d Schmiede

1086 forge; forge fire
[for]
f feu de forge; forge
e fragua; fuego de fragua
i forgia; fucina
n smidsvuur
d Schmiedefeuer; Schmiedeherd

1087 forge scale
[for]
f paille de fer; scorie de forge
e batiduras de forja
i scaglie di fucinazione
n hamerslag
d Zunder-Hammerschlag

1088 forging
[for]
f forgeage
e forjadura
i fucinatura
n smeden
d Schmieden

1089 forging machine
[for/mec]
f presse à forger
e prensa de forjar
i pressa per forgiare
n smeedmachine; smeedpers
d Schmiedepresse

1090 form grinding; plunge grinding; profile grinding
[gri]
f rectification de profils
e rectificación de perfiles

i rettifica di profili
n profielslijpen
d Profilschleifen

1091 forming cutter; form tool
[fre]
f fraise à profiler
e fresa de perfilar
i fresa da profilare
n profielfrees
d Profilfräser

1092 formwork hammer; shuttering hammer
[bui/ham]
f marteau de coffrage
e martillo de encofrado
i martello di cassaforma
n bekistingshamer
d Schalungshammer

1093 formwork plane; shuttering plane
[bui/pla]
f rabot de coffrage
e cepillo de encofrado
i pialla di cassaforma
n bekistingsschaaf; betonschaaf
d Betonschalungshobel

1094 foundation bolt; jag bolt; rag bolt
[fas/iwa]
f boulon de fondation
e perno de cimentación
i bullone di fondazione
n fundatiebout
d Fundamentbolzen

1095 foundry
[fou]
f fonderie
e fundería
i fonderia
n gieterij
d Giesserei

1096 foundry ladle
[fou]
f cuiller à couler; cuiller de coulée
e cazo de colada
i cucchiaione da fonderia
n gietpan
d Giesskelle; Giesslöffel; Giesspfanne

* **foundry ladle → 429**

1097 four-jaw chuck
[cpt/dri]
f mandrin à quatre mors
e mandril de cuatro garras; plato tetrogarras
i mandrino a quattro griffe
n boorhouder met vier bekken
d Vierbacken-Bohrfutter

1098 four-tined hayfork
[agr]
f fourche à foin à quatre dents
e horca de cuatro puntas para heno
i forcone a quattro denti per fieno
n hengst; viertandshooivork
d vierzinkige Heugabel

1099 four-way tool post
[cpt/fre/lat]
f porte-outil
e portaherramienta de cuatro vías
i portautensile a quattro vie
n vierbaansgereedschapshouder
d Vierwegewerkzeughalter

1100 foxtail wedging
[car/joi/woo]
f assemblage à contreclavette
e empalme con contrachaveta
i collegamento a controchiavetta
n tegenspieverbinding
d Gegenkeilverbindung

1101 frame level; mason's level
[mas/mea]
f niveau de maçon
e nivel de albañil
i livella di muratore
n metselaarswaterpas
d Maurerwasserwaage

1102 frame saw; span saw
[saw]
f scie à monture; scie à châssis
e sierra de ballesta; sierra de bastidor
i sega a telaio; sega intelaiata
n spanzaag
d Spannsäge

1103 frame saw spigot
[cpt/saw]
f chaperon de scie
e espiga de bastidor
i perno di telaio
n dwarspen
d Sägeangel

1104 framing chisel; heading chisel; joiner's mortise chisel; mortise chisel
[car/chi]
f bédane; ciseau à mortaiser
e cincel mortajador; escoplo grueso
i bedano

n schietbeitel
d Lochbeitel

* **framing square → 417**

1105 **fret saw** *GB*; **scroll saw** *US*
[saw]
f scie à chantourner; scie à découper
e sierra de calados; sierra de contornear
i gattuccio; sega a volgere
n figuurzaag; fretzaag; kromzaag
d Dekupiersäge; Laubsäge; Schweifsäge

1106 **fret-saw blade** *GB*; **scroll-saw blade** *US*
[saw]
f lame à découper
e hoja de sierra de calados
i lama del gattuccio; lama della sega a trafori; lama della sega a volgere
n figuurzaagblad
d Dekupiersägeblatt; Laubsägeblatt; Schweifsägeblatt

1107 **fret-saw frame** *GB*; **scroll-saw frame** *US*
[saw]
f arc de la scie à chanfreiner; porte-scie à découper
e arco de la sierra de calados
i archetto del gattuccio; archetto della sega a trafori; archetto della sega a volgere
n figuurzaagbeugel; figuurzaagboog
d Dekupiersägebogen; Laubsägebogen; Schweifsägebogen

1108 **friction saw**
[saw]
f scie râpante; scie sans denture
e sierra de fricción
i sega ad attrito
n smeltzaag
d Reibsäge

1109 **friction saw blade**
[cpt/saw]
f lame de scie râpante; lame de scie sans denture
e hoja de sierra de fricción
i lama di sega ad attrito
n smeltzaagblad
d Reibsägeblatt

1110 **froe; frow; riving axe**
[axe/foy]
f coin à fendre le bois
e cuña de partir madera
i cuneo da fendere legno
n splijtwig
d Spaltkeil

1111 **frog**
[cpt/pla]
f chariot
e carro
i carrello
n kikker
d Frosch

1112 **frog adjusting screw**
[cpt/pla]
f vis de réglage du chariot
e ajustador del carro
i vite di regolazione del carrello
n kikkerstelschroef
d Einstellschraube der Messerplatte (Frosch)

1113 **front**
[pla]
f front
e frente
i fronte
n voorkant
d Stirnseite

1114 **front bench hook**
[car/cpt]
f presse de devant d'établi
e tope frontal del banco
i cane frontale del banco
n voorklauw; voortang
d Vorderzange

1115 **front bit gouge; spoon bit gouge**
[chi]
f gouge cuillère
e gubia de cuchara
i sgorbia in forma di cucchiaio
n lepelguts
d Löffelhohlbeitel

* **front cutting nippers → 894**

* **frow → 1110**

1116 **fruit gatherer**
[agr/hor]
f cueille-fruit
e cogedor de frutas
i coglifrutta
n vruchteplukker
d Obstpflücker

1117 **frying pan**
[hou]
f poêle à frite; poêle à queue
e sartén con mango
i padella

n bakpan; koekepan
d Kuchenpfanne; Stielpfanne

1118 fuse pliers; fuse tongs
[pli]
f pince pour fusibles
e tenaza de seguras
i pinza per interni
n zekeringentang
d Sicherungsringzangen

G

1119 **gab**
[chi/sto]
f ciseau pour pierre des Ardennes
e útil para labrar sillar
i utensile a punta per lavorare pietre dure
n hardsteenbeitel
d Hartsteinmeissel

1120 **gablock; gavelock; iron crowbar**
[too]
f verdillon en fer
e palanca de hierro
i palanchino in ferro
n ijzeren koevoet
d eisernes Brecheisen

* **gablock → 657**

1121 **gad**
[chi]
f ciseau pointu
e escoplo de punta
i scalpello a punta
n puntbeitel
d Spitzmeissel

1122 **gad**
[min/sto]
f coin en acier
e cuña de cantero; cuña en acero
i cuneo in acciaio; cuneo per frantumare la roccia
n stalen wig
d Stahlkeil

1123 **gagger** *US*; **lifter** *GB*
[fou]
f tirant d'enlèvement; tirette
e repartidor de esfuerzos
i tirante di estrazione
n uitneemgereedschap
d Aushebeband

* **gambrel → 397**

1124 **gang drilling machine; multiple drilling machine**
[dri]
f perceuse à broches multiples
e taladro de brocas en hilero; taladro multihusillo; taladro múltiple
i trapano a punte multiple
n boormachine met meerdere spillen
d Mehrspindelbohrmaschine; Reihenbohrmaschine; Vielspindelbohrmaschine

1125 **gang mill** *US*; **gang saw** *GB*; **reciprocating saws** *GB*; **vertical log frame** *GB*
[foy/saw]
f scies à cadre à lames multiples
e sierras alternativas de hojas múltiples
i seghe alternative a telaio
n meervoudige raamzaag
d Vollglattersäge

1126 **gang miller**
[fre]
f fraise combinée
e tren de fresas
i fresa multipla
n combinatiefrees
d Gruppenfräser

1127 **gang milling**
[fre]
f fraisage combiné
e fresado con tren de fresas
i fresatura multipla
n combinatiefrezen
d Gruppenfräsen

1128 **gang mo(u)ld**
[fou]
f moule multiple
e molde múltiple
i forma multipla
n veelvoudige gietvorm
d Mehrfachgussform

1129 **gang saw; multiple saw; nest saw**
[saw]
f scie multilames
e sierra múltiple
i sega a lame intercambiabili
n meerbladige zaag; nestzaag
d Nestsäge

* **gang saw** *GB* **→ 1125**

1130 **gang-saw frame; gate**
[cpt/saw]
f cadre porte-lames; châssis porte-lames
e bastidor de la sierra alternativa
i telaio portaseghe
n zaagraam
d Gatter; Sägerahmen; Sägegestell

1131 **gang tool holder**
[fre]
f porte-outil à fraises multiples
e portaherramienta con fresas múltiples
i portautensile con frese multiple
n meervoudige freeshouder
d Mehrfachfräserträger

1132 garden cart; garden truck
[hor]
f brouette de jardin
e carretilla de jardinero
i carrello da giardino
n tuinkruiwagen
d Gartenschubkarren

1133 gardener's saw blade; pruning saw blade
[cpt/saw]
f lame jardinier
e hoja de sierra de podar
i lama di sega da portatore
n snoeizaagblad
d Gärtnersägeblatt

1134 garden hand syringe
[hor]
f seringue à main
e aspersor de mano
i spruzzatore a mano
n handspuit; straalpijp
d Gartenhandspritze

1135 garden hose
[hor]
f tuyau d'arrosage
e manguera de jardín
i tubo per annaffiare; tubo per irrigazione
n tuinslang
d Gartenschlauch

1136 garden rake
[hor]
f râteau de jardin
e rastrillo de jardín
i rastrello di giardino
n tuinhark
d Gartenrechen

1137 garden sprinkler
[hor]
f arroseur fixe
e aspersor rotativo
i irroratore girante
n vaste tuinsproeier
d Gartenregner

1138 garden trowel
[hor]
f transplantoir
e transplantador
i paletta da giardino
n tuinschopje; verplantschopje
d Erdschäufelchen; Pflanzkelle

* **garden truck → 1132**

1139 garnet hinge; strap hinge
[bui/joi]
f penture droite
e bisagra de ramal; pernio de T
i bandello diritto
n zware plaatduimheng
d Langband

1140 garnet paper
[gri/san]
f papier abrasif à poudre de grenat
e papel abrasivo revestido de polvo de granate
i carta vetrata a polvere di granato
n granaatpapier
d Granatpapier

1141 garnish bolt
[iwa]
f boulon décoratif
e perno decorativo
i bullone decorativo
n sierbout
d Zierbolzen

1142 gas-and-pressure-air burner
[bur]
f brûleur à gaz mixte
e quemador de gas mixto
i bruciatore a gas misto
n menggasbrander
d Mischgasbrenner

1143 gas-concrete saw
[saw]
f scie pour béton cellulaire
e sierra de hormigón celular
i sega per calcestruzzo cellulare
n gasbetonzaag
d Gasbetonsäge; Zellbetonsäge

1144 gas mask
[saf]
f masque antigaz
e careta antigas; máscara antigas
i maschera antigas
n gasmasker
d Gasmaske

1145 gas pipe tongs; gas pliers
[hyd]
f pince à gaz; pince serre-tubes
e alicate(s) de gasista; tenaza de fontanero
i giratubi; pinza per tubi
n gasfitterstang; gaspijptang; pijptang
d Gasrohrzange

1146 gas stocks and dies
[hyd]

f filière pour tubes à gaz
e terraja para tubos de gas
i filiera per filettare tubi di gas
n gastappen
d Gasrohrgewindeschneider; Gasrohr-schneidkluppe

1147 gate
[saw]
f cadre
e bastidor; marco
i telaio
n freem
d Sägegestell; Sägerahmen

* **gate → 1130**

1148 gate stick *GB*; **runner stick** *GB*; **sprue stick** *US*
[fou]
f broche de coulée
e canilla de bebedero
i bastone di colata
n ingietstok
d Eingussstock

1149 ga(u)ge
[mea]
f calibre; jauge
e calibre; galga
i calibro
n ijkmaat; kaliber
d Eichmass; Kaliber; Lehre

* **ga(u)ge box → 169**

1150 ga(u)ge pot
[mas/mea]
f caisse d'essai
e caja de ensayo
i cassetta di esame
n proefbakje
d Prüfkästchen

1151 ga(u)ge rod
[mea/tra]
f barre de sonde; velte
e varilla de sonda; varilla graduada
i sonda livello olio
n peilstok
d Peilstab

1152 ga(u)ging rod
[coo/mea]
f velte pour juger les tonneaux
e aforador diagonal; diagonal aforadora
i asta per misurare le botti
n boompasser
d Eichstab; Küferkrummesser

1153 ga(u)ging trowel; pointed trowel
[mas]
f truelle langue de chat; truelle maçon pointue; truelle pointue
e paletín de punta aguda
i cazzuola a punta
n pleistertroffel; punttroffel
d Katzenzungenkelle; Spitzkelle

* **gavel → 96, 461**

* **gavelock → 657, 1120**

* **G clamp → 443**

* **G cramp → 443**

1154 gear
[eng/gen]
f engrenage
e engranaje
i ingranaggio
n drijfwerk
d Getriebe; Treibwerk

1155 gear cutter; gear milling cutter
[eng/fre/gen]
f fraise à module; fraise à tailler les engrenages
e fresa de forma del diente; fresa de paso módulo
i fresa a modulo
n modulfrees
d Modulfräser; Zahnformfräser

* **gear frame brace → 619**

1156 gents saw
[saw]
f scie à dos pour ébénistes
e sierra fina
i sega a dorso per ebaniste
n roffelzaag
d Feinsäge; Zinkensäge

1157 geologist's hammer
[ham]
f marteau de géologue
e martillo de geólogo
i martello di geologo
n geologenhamer
d Geologenhammer

1158 gib-headed key
[fas/iwa]
f clavette à mentonnet; clavette à talon; contre-clavette
e contraclavija; chaveta con talón
i chiavetta a nasello; chiavetta a naso

n borgpen
d Nasenkeil

1159 gilder's brush
[art/dec]
f pinceau pour dorure
e pincel de dorador
i pennello per dorare
n verguldkwastje; verguldpenseel
d Anschiesser; Vergoldepinsel

1160 gimlet; wimble
[dri]
f vrille
e barrena de mango
i succhiello
n fretboor
d Nagelbohrer

1161 gimp needle; gymp needle
[lea]
f aiguille à garnir
e aguja para las orlas
i ago da rifinatura
n garneernaald
d Garniernadel

1162 gin wheel; pulley; sheave
[cpt/tra]
f poulie
e motón; polea
i puleggia
n katrol
d Rolle; Scheibe

1163 gland
[cpt/eng]
f presse-étoupe
e prensaestopas
i boccola di tenuta; premistoppa; pressatrecce
n pakkingbus
d Stopfbüchse

1164 gland bolts
[cpt/eng]
f boulons de la presse-étoupe
e pernos de la prensaestopas
i bulloni della premistoppa
n pakkingbusbouten
d Stopfbüchsenbolzen

1165 glassblower's tongs
[gla]
f pince-précelles
e tenaza de soplador
i pinza da vetraio
n glasblazerstang
d Zwackeisen

1166 glass crucible; glass pot
[gla]
f creuset
e crisol
i crogiolo
n glaskroes
d Glastiegel

1167 glass cutter
[cut/gla]
f coupe-verre
e tallador de vidrio
i tagliavetri
n glassnijder
d Glasschneider

1168 glass drill
[gla]
f perce-verre
e taladro para vidrio
i trapano di vetro
n glasboor
d Glasbohrer

1169 glass dropper
[che/mea]
f compte-gouttes
e cuentagotas
i contagocce
n droppelbuisje
d Tropfröhrchen

* **glass paper → 1062**

* **glass pot → 1166**

1170 glazier's chisel
[chi/gla]
f ciseau de vitrier
e cincel de vidriero
i scalpello di vetraio
n glazenmakersbeitel
d Glaserbeitel

1171 glazier's grooving plane
[gla/pla]
f feuilleret de vitrier
e guillame de vidriero
i incorsatoio di vetraio
n glassponningsschaaf
d Glaserhobel

1172 glazier's hammer
[gla/ham]
f marteau de vitrier
e martillo de vidriero
i martello di vetraio
n glashamer; glazenmakershamer
d Glaserhammer

1173 **glazier's pincers; tower pincers**
[gla/pli]
f pince à gruger; pince pour vitrier
e tenaza de vidriero; tenaza para quebrar vidrio
i tenaglie di vetraio
n glasbreektang
d Glasbrechzange; Glaserzange

1174 **glazier's T square**
[gla/mea]
f équerre coupe-verre en T du vitrier
e escuadra en T de vidriero
i squadra in T del vetraio
n glazenmakershaak
d T-förmiger Glaserschneidewinkel

1175 **glue-bonded abrasive**
[gri/san]
f abrasif à agglomérant colle
e abrasivo ligado con cola
i abrasivo ad impasto di colle
n met lijm gebonden slijpmiddel
d leimgebundenes Schleifmittel

1176 **glue brush**
[che/joi]
f brosse à coller
e cepillo para cola
i spazzola per colla
n lijmborstel
d Leimbürste

1177 **go and not go ga(u)ge; go, not-go ga(u)ge**
[mea]
f calibre entre-n'entre pas
e calibre de máximo y mínimo
i calibro passa non passa
n kaliber met goedkeur- en afkeurzijde
d Prüflehre mit Gut- und Ausschusseite

1178 **go-devil; porcupine**
[cle]
f hérisson de ramonage
e raspatubos
i raschiatore per tubi
n buiskrabber; buisschrapper
d Rohrreiniger; Rohrreinigungs-Spiralwelle; Rohrwischer

* **gold-and-silversmith's hammer → 1181**

* **goldbeater's pan → 211**

* **goldbeater's skin → 215**

1179 **gold beating hammer**
[gas/ham/met]
f marteau pour le battement de l'or en feuilles
e martillo batidor del oro
i martello per battitura dell'oro
n boeshamer; goudslagerijhamer
d Blattgoldschlägerhammer

1180 **goldsmith's beck iron**
[gas]
f tas plat d'orfèvre
e tas de orfebre
i tassello di orafo
n goudsmidstas
d Goldschmiedetasso; Knopfamboss

* **goldsmith's file → 987**

1181 **goldsmith's hammer; gold-and-silversmith's hammer**
[gas/ham/met]
f marteau d'orfèvre
e martillo de orfebre; martillo de orifice
i martello di orafo
n goudsmidshamer
d Goldschmiedehammer

* **go, not-go ga(u)ge → 1177**

1182 **gouge bit; shell bit**
[dri/woo]
f mèche à cuiller
e broca de cuchara
i punta a cucchiaio
n lepelboor
d Löffelbohrer; Parallelbohrer

1183 **gouges**
[chi]
f gouges
e gubias
i sgorbie
n gutsen; holbeitels
d Hohlbeitel

1184 **gouge slip; oilstone slip; slip stone**
[cpt/sha]
f pierre à affiler en forme de gouge
e piedra de amolar en forma de gubia
i pietra da affilare in forma di sgorbia
n gutsensteen
d Hohlbeitelstein

1185 **grab; grab-bucket**
[cpt/dre]
f grappin à deux mâchoires
e pala de doble concha
i benna a due valve; benna a trave; benna-draga

n grijpemmer
d Greifer; Zweiseitgreifer

1186 **grade**
[gri/mea]
f indice de dureté
e signo indicador de dureza
i indice di durezza
n hardheidsletter
d Härtebuchstabe

1187 **grade peg; grade post; gradient peg; gradient post**
[mea/sur]
f piquet
e piquete
i picchetto
n piket
d Absteckpfahl; Pikettpfahl

1188 **grading instrument; gradiograph**
[mea/sur]
f clinomètre
e clinómetro
i clinometro
n hellingmeter
d Gefällmesser

1189 **gradiometer; gradometer**
[mea/sur]
f clisimètre; niveau de pentes
e nivel de pendientes
i livella di pendenti
n hellingsgraadmeter
d Neigungsgradmesser

1190 **grafting knife**
[foy/hor]
f écussonnoir; entoir; greffoir
e abridor; cuchillo engeridor
i coltello da innesto; innestatoio
n entmesje; grifmesje
d Propfmesser

1191 **graining brush**
[dec]
f spalter; veinette
e cepillo de granelar
i pennello per marezzi
n aanlegkwastje; spalter
d Fladerpinsel

* **graining comb** → **560**

* **grain shovel** *US* → **626**

1192 **grape scissors**
[agr/hor/hou]
f ciseaux à raisins
e tijeras de uvas
i cesoie ad uve
n druiveschaar
d Traubenschere

1193 **grapevine jointer**
[mas]
f fer à joint côtelé
e paletín rayado
i ferro per giunti con costole
n voegspijker voor gesneden voeg
d Fugenschneider

1194 **graphite spray**
[lub]
f pinceau à graphite
e jeringa de grafito
i spruzzatore a grafite
n grafietspuitje
d Graphitspritze

1195 **grapnel; grappling hook; grappel**
[nav]
f grappin
e arpón de abordaje
i rampicone d'abbordaggio
n enterhaak
d Enterhaken

1196 **grass mower; lawn mower**
[hor]
f tondeuse à gazon
e cortacésped
i tosatrice da prato
n gazonmaaier; grasmaaimachine
d Rasenmäher

1197 **grass rake**
[hor]
f râteau d'herbe
e rastrillo para hierba
i rastrello per erba
n grashark
d Grasrechen

1198 **grass shears; lawn shears**
[hor]
f ciseaux à gazon
e tijeras para herba
i forbici da prato
n gazonschaar; grasschaar
d Grasschere; Rasenschere

1199 **grater**
[hou]
f râpe
e rallador; rallo
i grattugia

n keukenrasp; rasp
d Reibe

1200 **gravel rake; pebble rake; shingle rake**
[hor]
f râteau à gravier; râteau à gravillon
e rastrillo de grava
i rastrello di ghiaia
n grindhark; grinthark; kiezelhark
d Kiesharke; Kiesrechen

1201 **gravel roller; pebble roller; shingle roller**
[hor]
f rouleau à gravier; rouleau à gravillon
e rulo de grava
i rullo di ghiaia
n grindrol; grintrol; kiezelrol
d Kieswalze

1202 **graver's file**
[art/far]
f lime de graveur
e lima de grabador
i lima di tagliatore
n graveerdersvijl; graveurvijl
d Graveurfeile

* **graving tool → 368**

* **gravity drop hammer → 268**

1203 **grease cup**
[lub]
f graisseur
e engrasador
i ingrassatore
n vetbeker
d Schmierbecher

1204 **grease gun**
[gen/lub]
f pompe à graisse
e bomba de engrase a mano; jeringa de engrase
i ingrassatore a pressione
n smeerpomp; vetpomp
d Schmierkompressor

1205 **grinder dressing brick; plain rubbing brick**
[gri/san]
f brique-dresseur de meule
e piedra rectificadora
i pietra equilibratrice
n schuursteen
d Abrichtstein aus Karborund

1206 **grinder dressing diabolo**
[gri/san]
f diabolo de dressage de meule
e diábolo rectificador de muela
i diabolo equilibratore di mola
n opwerktol
d Abrichtkreisel

1207 **grinders**
[gri]
f meuleuses
e esmeriladoras
i smerigliatrici
n slijpmachines
d Schleifer

1208 **grinders**
[pri]
f meules; meuletons en grès
e muelas de piedra; muelas redondas de piedra
i molazze; pietre giranti
n koldermolens; pulpers
d Kollergänge; Läufersteine

* **grinder scouring serrated wheels → 10**

1209 **grinding apparatus**
[gri]
f appareil à meuler
e aparato esmerilador
i apparecchio smerigliatore
n slijpinrichting
d Geradeschleifgerät; Schleifapparat; Schleifeinrichtung

1210 **grinding steel; sharpening steel**
[gri/hou/nut/sha]
f affiloir; fusil
e chaira; eslabón
i affilatoio
n aanzetstaal; wetstaal
d Wetzstahl

1211 **grinding wheel flange**
[cpt/gri]
f flasque de meule
e reborde de muela
i flangia di mola
n slijpschijfflens
d Schleifscheibenflansch

1212 **grinding wheels**
[gri/san]
f meules
e muelas
i mole
n slijpschijven
d Schleifscheiben

1213 grommet *US*; **grummet** *GB*
[fas/gen]
f rondelle; virole
e arandela
i riparella; rondella
n sluitring
d Dichtungsring; Dichtungsscheibe

1214 grommet *US*; **grummet** *GB*
[fas/gen]
f bague d'étoupe; étoupage en chanvre
e empaquetadura de cáñamo
i guarnitura di canapa
n henneppakking
d Hanfdichtung; Hanfliderung; Hanfpackung

1215 groove and tongue joint; tongue and groove joint
[car/joi/woo]
f assemblage à rainure et languette
e ensambladura de ranura y lengüeta
i incastro a maschio e femmina
n veer- en groefverbinding
d Spund- und Nutverbindung

* **groove bit → 473**

* **groove cutter → 469**

* **grooved bit → 242, 793**

1216 grooved design block
[gas]
f dé à rainures
e dado ranurado
i tassello a scanalatura
n gleuvenank; sleuvenank
d Charnieramboss; Riefenanke; Rillenanke

1217 groove milling tool; slot milling tool; slot cutter
[fre]
f fraise à rainurer
e fresa de ranurar
i fresa per scanalare
n gleuffrees; sleuffrees
d Nutenfräser

1218 groove plane; grooving plane
[car/pla/woo]
f bouvet femelle
e cepillo acanalador
i sponderuola a bastone
n groefschaaf; moerploeg
d Nuthobel

1219 grooving
[fre]
f rainurage
e ranurado
i scanalatura
n groeven
d Nutenbildung

1220 grooving file
[far]
f lime à rainurer
e lima para ranurar
i lima per scanalare
n gleufvijl
d Nutenfeile

1221 grooving knife
[lea]
f rénette
e abreojetes
i lesina
n formeerijzer; groefmes
d Ritzmesser

* **grooving plane → 1218**

1222 grooving saw
[saw]
f scie à rainurer
e sierra ranuradora
i sega per scanalare
n groefzaag
d Nutsäge

1223 grooving tool; parting tool
[car/chi/lat]
f ciseau de tourneur
e escoplo de aplanar
i scalpello da tornitore
n afsteekbeitel
d Abstechstahl; Grabstichel; Plattenstahl

* **grooving tool → 354**

1224 ground auger
[dri/min]
f sonde; trépan
e barrena terrera
i trivella per pozzi artesiani
n aardboor; grondboor
d Erdbohrer

1225 ground bevel
[chi/cpt]
f biseau
e bisel
i smusso
n vouw
d Fase

1226 **grozing iron**
[gla]
f grésoir; grugeoir
e grujidor
i grisatoio
n gruisijzer
d Glasbrecheisen

1227 **grub axe**
[agr/hor]
f extirpateur
e extirpador
i estirpatore
n rooihak
d Rodehacke; Wurzelhacke

* **grubber** *US* **→ 663**

1228 **grub saw**
[saw]
f sciotte à dents
e sierra de marmolista de mano
i sega a lama per marmo
n marmerbladzaag
d Marmorblattsäge

1229 **grub screw**
[fas/scr]
f vis de pointage
e espárrago; tornillo de sujeción
i gambo filettato; perno filettato; vite di pressione
n stelschroef; tapbout
d Stiftschraube

* **grummet** *GB* **→ 1213, 1214**

1230 **guide fence**
[cpt/pla]
f butée transversale
e tope limitador transversal
i arresto trasversale
n zijgeleider
d Queranschlag

1231 **guide pulley; idler pulley**
[tra]
f galet-guide; galet de renvoi; poulie-guide
e polea de desviación; polea de guía; polea directriz
i galoppino; puleggia di guida
n leirol
d Führungsrolle; Leitrolle; Umlenkrolle

1232 **guiding stick; maulstick**
[dec]
f appui-main
e apoyamano; tiento
i appoggiamano; bacchetto dei pittori
n schildersstok
d Malerstock

1233 **guillotin shears**
[she]
f cisaille à guillotine
e tijeras de guillotina
i cesoie a ghigliottina
n guillotineschaar
d Rahmenschere

* **gullet → 340**

* **gullet saw → 341**

* **gullet tooth → 340**

1234 **gun; template; templet**
[fas]
f traçoir
e trazador
i tracciatore
n afschrijfblokje; afschrijfplankje
d Abschreibbrett

* **gun drill → 718**

1235 **gurlet**
[ham/min]
f grelet
e martillo para picar
i martello per piccare
n punthamer
d Spitzhammer

* **gymp needle → 1161**

1236 **gyratory crusher**
[eng/nut]
f concasseur giratoire
e machacadora giratoria; quebrantadora giratoria
i frantoio rotante
n roterende breker
d Kreiselbrecher

H

1237 **hack; pick**
[axe/gen/sto]
f pic
e pico
i piccone
n houweel
d Spitzhacke

1238 **hacking knife; hacking-out knife**
[cle/dec/gla]
f couteau à démastiquer; hachette de vitrier
e cuchilla de vidriero; cuchilla para alejar la masilla
i coltello per rimuovere il mastice
n schildershakmes
d Kittentfernungsmesser

1239 **hackle**
[agr/tex]
f peigne; séran
e rastrillo
i pettine
n hekel
d Hechel

1240 **hackle block; hackle stock**
[agr/tex]
f chaise à sérancer
e banco de rastrillar
i cavalletto del pettinatore
n hekelstoel
d Hechelblock

1241 **hacksaw**
[saw]
f scie à archet à métaux
e sierra de arco para metales
i sega per metalli ad arco
n beugelmetaalzaag
d Bügelmetallsäge

1242 **hair compasses**
[mea]
f compas à pointes à charnière, guide et vis d'arrêt
e compás de puntas con bisagra, guía y tornillo de ajuste
i compasso a punte con cerniera, guida e vite di fissaggio
n steekpasser met scharnier en stelschroef
d Spitzzirkel mit Nietscharnier, Bogen und Stellschraube

1243 **hair pencil**
[art/dec]
f pinceau en poil de chameau
e pincel en pelo de camello
i pennello di pelo di cammello
n kameelhaarpenseel
d Kamelhaarpinsel

1244 **hair sieve**
[agr/hou]
f tamis de crin
e codazo capilar; tamiz capilar
i setaccio di crine
n haarzeef; haren zeef
d Haarsieb

1245 **half-back saw**
[saw]
f scie à demi dos
e sierra de medio dorso
i sega a mezzo dorso
n zaag met halve rug
d Säge mit halbem Rücken

1246 **half lap-joint; halving**
[car/joi/woo]
f assemblage à mi-bois
e ensambladura a media madera
i giunto a mezzo legno
n halfhoutverbinding
d Halbholzverbindung

1247 **half-rip saw**
[saw]
f scie de long à denture fine
e sierra al hilo con dentadura fina
i sega per il lungo a dentatura fina
n schulpzaag met fijne vertanding
d Brettsäge mit feiner Zahnung

1248 **half-round chisel**
[chi/woo]
f ciseau demirond
e cincel media caña
i scalpello mezzotondo
n halfronde beitel
d halbrunder Beitel

1249 **half-round file**
[far]
f demironde; lime demironde
e lima semirredonda
i lima mezzotonda
n halfronde vijl
d Halbrundfeile

1250 **half-round profile plane**
[pla]
f rabot à boudins
e cepillo de perfiles de media caña
i pialla di profili mezzotondi

n beulingschaaf
d Halbrundprofilhobel

* **half-round screw → 381**

1251 **half-round sharpening stone**
[gri/sha]
f pierre à affûter demironde
e piedra de amolar media caña
i pietra da affilare mezzotonda
n halfronde aanzetsteen
d Halbrundschleiffeile

1252 **half-round spokeshave; hollow spokeshave**
[car/pla/woo]
f vastringue à semelle concave
e bastren hueco
i pialletto concavo per sagomare
n holle spookschaaf
d hohler Schabhobel

1253 **half-smooth cut**
[far]
f taille demidouce
e picadura semidulce
i taglio mezzodolce
n halfzoete kap
d halbschlichter Hieb; Hieb 2

1254 **half-smooth file**
[far]
f lime demidouce
e lima semidulce
i lima mezzodolce
n halfzoetvijl
d Halbschlichtfeile

* **halving → 1246**

1255 **hames block; horse collar block**
[fas/iwa]
f forme pour collier de cheval
e horma para colleras
i forma da collare
n gareelblok
d Kummetstock; Kumtstock

1256 **hames nail; horse collar nail**
[fas/iwa]
f clou de collier de cheval
e clavo de horcate
i chiodo per il corpo del collare
n gareelnagel; haamnagel
d Kummetnagel; Kumtnagel

1257 **hammer axe**
[axe/ham]
f hachette à marteau
e hacha con peto; hacha-martillo
i malatesta; scure a martello
n hamerbijl
d Hammerbeil

1258 **hammer bell; hammer head**
[cpt/ham]
f tête de marteau
e cabeza de martillo
i testa di martello
n hamerkop
d Hammerkopf

1259 **hammer crusher; impact crusher**
[ham/sto]
f broyeur à mâchoires
e machacadora; trituradora a mordazas
i frantoio a martelli
n kaakbreker
d Backenbrecher

1260 **hammer drill**
[dri/ham]
f marteau perforateur; perforatrice à percussion
e martillo perforador; perforadora de percusión
i martello perforatore a percussione
n boorhamer
d Bohrhammer

* **hammer-drive screw → 819**

1261 **hammer eye; hammer ring**
[cpt/ham]
f œil du marteau
e ojo de martillo
i occhio del martello
n hamergat; hamerhuis
d Hammerauge; Hammerloch; Hammerohr

* **hammer face → 931**

1262 **hammer grip**
[cpt/ham]
f poignée de marteau
e empuñadura de martillo
i impugnatura di martello
n hamerhandgreep
d Hammerhandgriff

1263 **hammer handle; hammer helve; hammer shaft**
[cpt/ham]
f manche du marteau
e mango del martillo
i manico del martello
n hamersteel
d Hammerstiel

1264 **hammer head**
[cpt/ham]
f masse tombante
e martillo pilón
i mazza battente
n beer
d Bär

* **hammer head → 1258**

* **hammer helve → 1263**

1265 **hammerman; smith's mate**
[for]
f dauteur
e ayudante de herrero; batidor
i battimazza; garzone di fabbro
n smidsknecht; voorslaander
d Schmiedegeselle; Zuschläger

1266 **hammer pane; hammer peen; hammer pein**
[cpt/ham]
f panne du marteau
e peña del martillo
i penna del martello
n hamerpen
d Hammerfinne; Hammerpinne

* **hammer ring → 1261**

1267 **hammers**
[gen/ham]
f marteaux
e martillos
i martelli
n hamers
d Hämmer

1268 **hammer scale**
[for]
f mâchefer; paille
e batiduras; escoria de hierro batido
i scaglie di ferro
n hamerslag
d Hammerschlag; Schmiedesinter

* **hammer shaft → 1263**

1269 **hammer wedge**
[cpt/ham]
f coin du manche de marteau
e cuña del mango de martillo
i cuneo del manico di martello
n hamerspie
d Hammerkeil; Krallenkeil

1270 **hand anvil**
[for]
f enclumeau
e yunque portátil
i incudine di banco
n handaambeeld
d Bankamboss; Handamboss

1271 **hand axe**
[axe]
f cognée; hachette
e hacheta; hachuela
i scure leggera
n handbijl
d Handbeil

1272 **hand chisel**
[chi]
f ciseau à main
e cincel de mano
i scalpello a mano
n handbeitel
d Handbeitel

1273 **hand cultivator** *GB*; **hand grubber** *GB*; **hand lister** *US*
[agr/hor]
f bêche semiautomatique
e laya semiautomática
i estirpatore
n cultivator
d Grubber; Kultivator

1274 **hand drill; hand drilling apparatus**
[dri]
f foreuse à main; perceuse à main
e barrena de mano; taladro a mano
i trapano a mano
n handboormachine
d Handbohrgerät; Handbohrmaschine

* **hand drilling hammer → 540**

1275 **hand file**
[far]
f lime à main
e lima de mano
i lima a mano
n blokvijl; handvijl
d Handfeile

1276 **hand grip**
[cpt/gen]
f poignée
e empuñadura
i impugnatura
n greep; handvat
d Griff; Handgriff

* **hand grubber** *GB* **→ 1273**

1277 **hand hammer**
[ham]
f frappeur; marteau à main
e martillo de mano
i martello a mano; martello di fucina
n handhamer
d Handhammer

1278 **hand harrow**
[agr/hor]
f herse à main
e grada de mano
i frangizolle a mano
n handeg(ge)
d Handegge

1279 **hand ladle; hand shank**
[fou]
f cuiller à couler; cuiller de coulée
e cazo; coladora; cuchara
i cazza; cucchiaio a mano; cucchiaio di colata
n gietlepel
d Giesslöffel

1280 **hand lamp; portable lamp**
[ele]
f lampe baladeuse
e lámpara portátil
i lampada portatile
n handlamp
d Handlampe

1281 **handle; helve; shaft**
[cpt/gen]
f manche
e astil; mango cabo
i manico
n steel
d Helm; Stiel

1282 **hand lead; hand sounder**
[mea/nav/sur]
f sonde à main
e escandallo de mano; sonda de mano
i scandaglio a mano
n handlood; licht peillood
d Handlot

1283 **handled**
[cpt/ham]
f emmanché
e con mango
i con manico
n met steel
d mit Stiel

1284 **handled with fixing straps**
[cpt/ham]
f à manche claveté; emmanché à clavettes
e con mango acuñado
i con manico a fasci di rinforzo
n met steel en zijplaten
d mit durchgenieteten Federn; mit verkeiltem Stiel

1285 **handle of a plane; horn of a plane; nose of a plane**
[cpt/pla]
f corne
e cuerno
i corno
n neus
d Nase

1286 **handle of a scythe; scythe handle**
[agr/cpt]
f manche de faux
e manija de guadaño
i impugnatura di falce
n zeiskruk
d Sensengriff

1287 **handle of a scythe; scythe handle**
[agr/hor]
f manche de faux
e astil de guadaña
i manico di falce
n zeisboom; zeisstok
d Wurf

1288 **handles of pliers**
[pli]
f branches d'une pince
e brazos de un alicate
i ganasce di una pinza
n benen van een tang
d Schenkel einer Zange

1289 **hand level**
[mea/sur]
f niveau portatif
e nivel portátil
i livella a mano
n handwaterpas
d Handwasserwaage

1290 **handle with top and end ferrule; heavy-duty handle; two-ferrule-handle**
[chi/cpt]
f manche à deux viroles
e mango de dos virolas
i manico a due virole
n hakhecht
d Zweizwingenhelf

* **hand lister** *US* → **1273**

1291 **hand log**
[mea/nav]
f loch à main
e corredera de mano
i solcometro a mano
n handlog
d Handlog

1292 **hand milling machine**
[fre]
f fraiseuse à main
e fresadora portátil
i fresatrice a mano
n handfrees
d Fräsmaschine mit Handbedienung

1293 **hand-rail plane**
[joi/pla]
f rabot pour main courante
e cepillo para pasamanos
i pialletto per corrimano
n leuningschaaf
d Handlaufhobel

* **hand-rail screw → 770**

1294 **hand rasp**
[far]
f râpe à main
e escofina de mano
i raspa a mano
n handrasp
d Handraspel

1295 **hand reamer**
[dri/fre/rea]
f alésoir à main
e escariador de mano
i alesatore a mano
n handruimer
d Handreibahle

1296 **hand rest**
[cpt/lat]
f support d'outils à main
e soporte de herramientas de mano
i lunetta del tornio; supporto di utensili di mano
n handgereedschapshouder
d Werkzeughalter

* **hand router → 724**

1297 **handsaw**
[saw]
f scie à main; scie égoïne
e serrucho de mano; sierra de mano
i sega a mano; segaccio a mano
n handzaag
d Fuchsschwanz; Handsäge

1298 **handsaw blade**
[cpt/saw]
f lame égoïne
e hoja de sierra de mano
i lama di sega a mano
n handzaagblad
d Fuchsschwanzblatt

1299 **handscrew; parallel cramp**
[joi]
f presse-joint
e sargento
i sergente
n meubelmakersklem
d Schraubzwinge

* **hand shank → 1279**

* **hand sounder → 1282**

1300 **hand sprayer**
[agr/hor]
f seringue à main
e jeringa de mano
i nebulizzatore a mano
n duimspuit
d Handzerstäuber

1301 **hand tap**
[dri/scr]
f taraud à main
e macho de mano de terrajar
i maschio sbozzatore a mano per filettare
n handtap
d Handgewindebohrer; Handschneidbohrer

1302 **hand tools**
[too]
f outillage à main; outils à main
e herramientas manuales
i arnesi a mano; utensileria a mano
n handgereedschap
d Handwerkzeuge

1303 **hand vice**
[cpt/vic]
f étau à main
e entenalla; tornillo de mano
i morsetto a mano
n handschroef
d Feilkloben

1304 **hardboard nail**
[fas/iwa]
f clou pour panneau dur
e clavo para tablas de fibra prensada

i chiodo per cartone pressato
n boardnagel
d Pressspannagel

1305 **hardboard plane**
[pla]
f rabot pour panneau dur
e cepillo para tablas de fibra prensada
i pialletto per cartone pressato
n boardschaaf
d Pressspanhobel

1306 **hardboard riveting apparatus**
[riv]
f appareil de rivetage du panneau dur
e aparato de remachado para cartón barnizado
i apparecchio di chiodatura per cartone pressato
n boardnietapparaat
d Pressspannietgerät

1307 **hardened teeth**
[cpt/saw]
f dents trempés
e dientes templados
i denti temperati
n geharde tanden
d gehärtete Zähne

1308 **hardness testing file**
[far/met]
f lime pour essais de dureté
e lima comprobadora de dureza
i lima per prova di durezza
n vijl voor hardheidsproeven
d Härteprüffeile

1309 **hard point teeth**
[cpt/saw]
f dents à pointes durcies
e dientes de puntas duras
i denti a punte temperate
n tanden met geharde punten
d Zähne mit gehärteten Spitzen

1310 **hardwood handle**
[cpt/gen]
f manche en bois dur
e mango de madera dura
i manico in legno duro
n hardhouten steel
d Hartholzheft

1311 **hardy hole**
[cpt/for]
f trou carré d'enclume
e agujero cuadrado de yunque
i foro quadrato d'incudine
n aambeeldgat; aanbeeldgat
d Ambossloch

1312 **harrow** *GB*; **soil pulverizer** *US*
[agr]
f herse
e grada
i erpice
n eg(ge)
d Egge

1313 **hatchet**
[axe]
f hachette
e hacheta
i accetta
n bijltje; kleine bijl
d Handbeil

1314 **hatchet iron**
[wel]
f fer à souder à marteau
e soldador tipo martillo
i saldatore a martello
n hamerbout; hamersoldeerbout
d Hammerlötkolben

* **hatchet stake → 75**

1315 **hawk; plasterer's hawk**
[mas]
f taloche de maçon
e paleta para argamasa; talocha de albañil
i frettazzo di muratore; sparviere
n pleisterbord; raapbord; spaarbord
d Mörtelbrett; Streichbrett

1316 **hay baler; pick-up baler**
[agr]
f ramasseuse-presse
e prensa recogedora
i pressa raccoglitrice
n hooiopraper
d Aufsammelpresse

* **hay elevator → 133**

1317 **hayfork**
[agr]
f fourche à faner
e bielda
i tridente
n hooivork
d Heugabel

1318 **hay grab hoist; hay pole**
[agr]
f déchargeur
e descargador

i scaricatore
n hooigrijper
d Heugreifer

1319 hay knife
[agr]
f coupe-foin
e cuchillo para heno
i coltello per fieno
n hooimes; hooisnijder
d Heumesser

* **hay loader → 135**

* **hay pole → 1318**

1320 hay rake
[agr]
f fauchet
e rastrillo para el heno
i rastrello per fieno
n hooihark
d Heuharke; Heurechen

1321 hay saw
[agr/saw]
f scie à foin
e sierra para heno
i sega per fieno
n hooizaag
d Heusäge

1322 haystack temperature feeler
[agr/mea]
f senseur de la température dans le tas de foin
e tentácula de la temperatura en el almiar de heno
i indicatore della temperatura nel mucchio di fieno
n hooiijzer; hooiroede
d Heutemperaturanzeiger

* **heading chisel → 1104**

1323 heading tools
[gen/too]
f outils à façonnement des têtes
e herramientas para formar cabezas
i arnesi per formare teste
n stuikgereedschap
d Stauchwerkzeuge

1324 headless bolt
[fas/iwa]
f boulon sans tête
e perno sin cabeza
i bullone senza testa
n koploze bout
d kopfloser Bolzen

1325 headless screw
[fas/scr]
f vis sans tête
e tornillo sin cabeza
i vite senza testa
n koploze schroef
d kopflose Schraube

1326 head of a bolt
[iwa]
f tête de boulon
e cabeza de perno
i testa di bullone
n kop van een bout
d Bolzenkopf; Kopf eines Bolzens

1327 head of a hammer
[ham]
f tête de marteau
e cabeza de martillo
i testa di martello
n hamerkop
d Hammerkopf

1328 head saw
[foy/saw]
f sciage du tronc; tronçonnage
e aserrado del tronco
i segatura del tronco
n zagen van de stam
d Sägen des Baumstammes

1329 headstock of a lathe; lathe headstock
[cpt/lat]
f poupée
e contrapunto
i testa motrice
n kop van draaibank
d Spindelkasten; Spindelstock

1330 headstock of a planing machine; planing machine headstock
[cpt/pla]
f porte-lame de raboteuse
e portahierra de acepilladora
i portascalpello di piallatrice
n beitelram van een schaafmachine
d Meisselhalter einer Hobelmaschine

1331 heath besom; heath broom
[cle/hou]
f balai à bruyère
e escoba de brezo
i granata di scopa
n heibezem
d Heidebesen

1332 **heath scrubber**
[cle/hou]
f brosse de chiendent
e limpiador de brezo
i spazzola di scopa
n heiboender
d Heidescheuerbürste

1333 **heavy-duty clamp**
[cla]
f presse à vis à fort débit
e prensa para trabajos fuertes
i sergente per servicio pesante
n zware ketelklem
d Schwerlastschraubzwinge

1334 **heavy-duty diagonal cutting pliers**
[pli]
f pince coupante diagonale à becs puissants
e alicate(s) de corte diagonal para trabajos pesados
i pinza per taglio diagonale per servizio pesante
n diagonaaltang voor zwaar werk
d Diagonalschneider mit kräftigen Backen

1335 **heavy-duty drilling machine**
[dri]
f perceuse à grand débit
e taladro de alta capacidad
i trapano di grande potenza
n boormachine met grote capaciteit
d Hochleistungsbohrmaschine

* **heavy-duty handle → 1290**

* **heavy-duty machine vice → 1041**

* **heavy-duty saw → 175**

1336 **heavy engineer's file**
[far]
f lime lourde d'atelier
e lima pesada de ajustador
i lima pesante di meccanico
n zware bankwerkersvijl
d schwere Werkstattfeile

1337 **hedge clipper; hedge shears**
[agr/hor/shel]
f taille-haies
e tijeras para recortar setos
i forbici per siepi
n heggeknipper; heggeschaar
d Heckenschere

1338 **heel-glazing iron**
[lea]
f polissoir de talon
e pata de cabra para tacones
i liscia per tacchi
n hakpolijstijzer
d Absatzglätteisen

1339 **heel knife; pairing knife**
[lea]
f tranchet recourbé
e tranchete curvo
i trincetto curvo
n krommes
d Krummesser

1340 **heel of a file**
[far]
f talon de lime
e talón de lima
i tallone di lima
n hals van een vijl
d Hals einer Feile

1341 **heel of a plane**
[cpt/pla]
f appui-main de rabot
e soporte manual de cepillo
i supporto manuale di pialla
n handsteun van een schaaf; hiel van een schaaf
d Handauflage eines Hobels

1342 **helical countersink**
[dri/rea]
f alésoir-fraise conique hélicoïdal
e avellanador cónico helicoidal
i accecatoio conico elicoidale
n spiraalverzinkfrees
d Spiralsenker

1343 **helical gears**
[cpt/eng]
f engrenages hélicoïdaux
e engrenajes en espiral
i ingranaggi elicoidali
n schroefwieloverbrengingen
d Spiralverzahnungsgetriebe

1344 **helical hinge**
[cpt/hou/iwa]
f penture pour porte va-et-vient
e bisagra de mampara
i cerniera per porte a vento
n doordraaiend scharnier
d Pendeltürband

1345 **helical spring**
[cpt/iwa]
f ressort en spirale
e resorte hélice; resorte helicoidal

i molla a spirale
n spiraalveer
d Spiralfeder

* **helicoidal saw → 1059**

* **helve → 1281**

1346 **helve ring**
[cpt/tool]
f anneau de manche
e anillo de astil; anillo de mango
i anello di manico
n heftring; steelring
d Heftring; Stielring

1347 **hermaphrodite cal(l)ipers; jenny cal(l)ipers; odd leg cal(l)ipers**
[mea]
f compas à élément de fixation
e calibre hermafrodito
i compasso calibro con elemento di fissazione
n passer met derde been
d Zirkel mit Stellbein

1348 **hexagonal broach**
[rea]
f broche pour trous hexagonaux
e brocha para agujeros hexagonales
i broccia per fori esagonali
n broots voor zeskante gaten
d Räumer für Sechskantlöcher

1349 **hexagonal head bolt; hex head bolt**
[fas/iwa]
f boulon à tête hexagonale
e perno de cabeza hexagonal
i bullone a testa esagonale
n bout met zeskante kop
d Bolzen mit Sechskantkopf

1350 **hexagonal head screw**
[iwa/scr]
f vis à tête hexagonale
e tornillo de cabeza hexagonal
i vite a testa esagonale
n schroef met zeskante kop
d Sechskantschraubenkopf

1351 **hexagonal key**
[cpt/spa]
f clé hexagonale; clé à six pans
e llave hexagonal
i chiave esagonale; chiave maschio per esagoni
n binnenzeskantsleutel; inbussleutel; stiftsleutel
d Sechskantschlüssel

1352 **hexagonal shank**
[cpt/gen/tool]
f queue hexagonale
e espiga hexagonal; mango hexagonal
i stelo esagonale
n zeskante kolf
d Sechskantschaft

1353 **hexagon dresser**
[cle/gri/sha]
f décrasse-meule hexagonale
e alisador hexagonal de muela
i lisciatore esagonale di mola
n zeskante slijpschijfscherper
d Sechskantschleifscheibenschärfer

1354 **hexagon nut**
[iwa/scr]
f écrou hexagonal
e tuerca hexagonal
i dado esagonale
n zeskante moer
d Sechskantmutter

* **hex head bolt → 1349**

1355 **high-frequency hardened teeth**
[cpt/saw]
f denture traitée par trempe haute fréquence
e dentadura de temple por corrientes de alta frecuencia
i dentatura temperata per correnti di alta frequenza
n hoogfrequent geharde tanden
d hochfrequenzgehärtete Zahnung

1356 **high leverage diagonal cutting pliers**
[pli]
f pince coupante diagonale à haute force de levier
e alicate(s) de corte diagonal de alta acción de una palanca
i pinza per taglio diagonale ad alto leveraggio
n diagonaaltang met grote hefboomwerking
d Hebeldiagonalschneider; Hebelseitenschneider

1357 **high pressure burner**
[bur]
f brûleur à haute pression
e quemador de alta presión
i bruciatore ad alta pressione
n hoge-drukbrander
d Hochdruckbrenner

1358 **high-speed drilling machine**
[dri]

f perceuse à grande vitesse; perceuse rapide
e taladro rápido
i trapano rapido
n boormachine met hoog toerental
d Schnellbohrmaschine

1359 high-speed expanding hand reamer
[dri/fre/rea]
f alésoir à main superexpansible
e escariador de mano de ajuste rápido
i alesatore a mano ad espansione rapida
n snelverstelbare ruimer
d Schnellverstellhandreibahle

1360 hinge
[fas/iwa]
f charnière
e bisagra
i cerniera
n scharnier
d Scharnier

1361 hinge framing chisel
[chi]
f bédane pour charnières; ciseau pour charnières
e cincel para bisagras
i bedano per cerniere; scalpello per cerniere
n knierbeitel; scharnierbeitel
d Fitschenbeitel

1362 hinge mortise bit; hinge mortise cutter
[car/fre/woo]
f fraise mortaiseuse
e fresa para mortajas de charnelas
i fresa per mortasare
n scharnierfrees
d Fensterwinkelfräser

1363 hob
[met]
f master; poinçon d'impression
e macho; punzón para estampación
i punzone di stampato
n patrijs
d Einsenkstempel; Leisten; Meister; Pfaffe

1364 hob; hobbing cutter
[fre/met]
f fraise-mère développante; fraise vis-mère
e fresa generadora; fresa madre
i fresa creatrice
n genererende frees
d Abwalzfräser

1365 hobbing foot; triple last
[lea]
f trépied de cordonnier
e horma triple; yunque de zapatero
i piede di ferro
n ijzeren drievoet
d Schusterdreifuss

* **hobby tools → 50**

1366 hobnail
[iwa/lea]
f caboche; clou à ferrer les souliers
e tachuela
i bulletta
n pummel; schoenspijker met dikke kop
d grober Schuhnagel

1367 hod
[mas/tra]
f auge; hotte de maçon
e cuezo
i calcinaio del muratore
n kalkbak; trog
d Kalktrog; Mörteltrog; Tragmulde

1368 hoe blade
[agr/hor]
f soc
e reja del aporcador
i lama della zappa
n hakblad; schoffelblad
d Häufelschar

1369 hoe handle
[agr/hor]
f manche de la houe; manche du hoyau
e mango de la arada
i manico della zappa
n haksteel
d Hackenstiel

1370 hoeing machine; steering hoe
[agr/hor]
f bineuse
e binadora
i sarchiatrice
n schoffelmachine
d Hackmaschine

1371 hoist
[gen/tra]
f monte-charge
e montacargas
i elevatore meccanico; montacarichi; paranco
n hijstoestel
d Aufzug

1372 hold-down bolt
[fas/iwa]

f boulon d'assujettissement
e perno de anclaje
i bullone di fissaggio al suolo
n verankeringsbout
d Verankerungsbolzen

1373 **holdfast**
[car/saf]
f crampon
e barrilete
i spranga
n klamp
d Spannklupe

1374 **holding screw**
[fas/scr]
f vis d'arrêt
e tornillo de ajuste
i vite di arresto
n stelschroef
d Stellschraube

1375 **holding up**
[iwa/riv]
f contrebouterolle
e sufridera
i controspinta
n tegenhouder
d Gegenhalter; Nietpfanne

1376 **hole boring cutter**
[dri/fre]
f fraise à aléser; fraise gouchette
e taladro de fresa
i trapano a fresa
n freesboor
d Bohrfräser

1377 **hole cutter; hole saw**
[car/saw/wool]
f scie-cloche
e sierra agujereadora
i sega a tazza
n gatzaag; zevengatenzaag
d Sägeglocke

1378 **hole cutter; washer cutter**
[iwa/met]
f outil pour former des rondelles
e fresa de barrenar; útil para cortar arandelas
i utensile per tagliare rondelle
n sluitringsnijder
d Unterlegscheibenschneider

1379 **hole cutting shears**
[met/she]
f cisaille perforatrice; cisaille pour trouer
e tijeras agujereadoras
i cesoia perforatrice
n gatenschaar
d Lochschere

1380 **hole ga(u)ge**
[mea]
f calibre d'alésage; calibre de perçage
e calibre de taladros; galga de agujeros
i calibro per fori
n gatkaliber
d Bohrungslehre; Lochlehre

* **hole saw → 1377**

1381 **hollow adze; rounding adze**
f herminette à fer recourbé; herminette à gouge
e azuela curva; azuela hueca
i ascia a filo ricurvo; ascia cava
n gebogen dissel
d Hohldexel

1382 **hollow chamfer**
[car/wool]
f arête creuse; cannelure
e caveto invertido
i smussatura cava
n holle groef
d Hohlkehle

1383 **hollow drill**
[dri]
f évidoir
e barrena hueca; broca hueca
i trapano cavo
n holboor
d Hohlbohrer

* **hollowing knife → 607**

1384 **hollowing plane**
[pla]
f rabot à gorge
e cepillo cóncavo; guillame hembra
i pialla concava; sponderuola a bastone
n creseerschaaf; hol; holschaaf
d Hohlkehlhobel

1385 **hollow-out** *v* **by filing**
[far]
f creuser à la lime
e ahondar con la lima
i limare per incavare
n uitvijlen
d ausfeilen

1386 **hollow punch**
[lea/per]
f emporte-pièce; perçoir

e punzón sacabocados
i punzone tubolare
n gaatstempel
d Locheisen

* **hollow punch → 198**

* **hollow spokeshave → 1252**

* **hollow-wall plug → 442**

1387 **hone; hone stone; whetstone**
[gri/sha]
f pierre à aiguiser; pierre à repasser
e piedra de afilar; piedra de amolar
i pietra da affilare
n aanzetsteen; wetsteen
d Abziehstein; Schleifstein

1388 **hone sheath; whetstone sheath**
[cpt/sha]
f coffin
e funda de la afiladera
i fodero della pietra da cote
n wetsteenvaatje
d Kumpf

* **hone stone → 1387**

1389 **honing**
[gri/sha]
f aiguisage
e afiladura; amoladura
i affilatura
n aanzetten; wetten
d Schleifen

1390 **honing machine**
[gri/sha]
f machine à roder
e bruñidora; máquina de alisar
i levigatrice; macchina per rettifica a lucido
n hoonmachine; uitslijpmachine
d Honmaschine

1391 **hook**
[fas/iwa]
f crochet
e gancho
i gancio
n haak
d Haken

1392 **hook**
[fas/iwa]
f clou à crochet
e escarpia
i arpione; ganghero
n duim
d Hakennagel; Hakenstift

1393 **hook bolt; hook screw**
[fas/iwa/scr]
f boulon à croc; boulon à crochet
e tornillo de gancho
i bullone a gancio
n haakbout
d Hakenschraube

1394 **hook ladder**
[fir/saf]
f échelle à crochets
e escalera de ganchos
i scala a gancio
n haakladder
d Hakenleiter

1395 **hook scraper**
[cle/iwa]
f ciseau à crochet; racloir à crochet
e rascador de gancho
i scalpello a gancio
n haakbeitel; haakijzer
d Hakenschaber

* **hook screw → 1393**

1396 **hook spanner**
[spa]
f clé à crochet; clé coudée
e llave acodada; llave de gancho
i chiave a gancio; chiave poligonale; chiave a stella
n armsleutel; haaksleutel
d Hakenschlüssel

1397 **hook spanner; nut runner; sickle spanner**
[scr/spa]
f clé à ergot pour écrous
e aprietatuercas mecánicas
i chiave a gancio
n moeraanzetsleutel
d Mutteranziehschlüssel

* **hook spanner → 497, 1022**

1398 **hoop driver; hoop setter**
[coo]
f batte; hutinet; mallet de tonnelier
e mazo para colocar aros
i mazzuolo per forzare cerchi
n kuipershamer
d Küfersetzhammer; Reifsetzer

1399 **hopper**
[agr/tra]

f boîte d'alimentation; trémie
e tolva
i tramoggia
n molentrechter; tremel
d Gosse; Haube; Kornkasten

1400 horizontal drilling machine
[dril]
f perceuse horizontale
e taladro horizontal
i trapano orizzontale
n horizontale boormachine
d Horizontalbohrmaschine

1401 horizontal miller; horizontal milling machine; horizontal spindle milling machine
[fre]
f fraiseuse horizontale
e fresadora horizontal
i fresatrice orizzontale
n horizontale freesmachine
d Wagerechtfräsmaschine

* **horn of a plane → 1285**

1402 horse brush
[cle]
f brosse de pansage
e bruza
i brusca
n paardeborstel
d Kardätsche

* **horse collar block → 1255**

* **horse collar nail → 1256**

1403 horse rasp
[far/for]
f carpe maréchal; lime-râpe maréchal
e escofina de herrero
i raspa di maniscalco
n hoefrasp
d Hufraspel

1404 horseshoe
[for]
f fer à cheval
e herradura
i ferro di cavallo
n hoefijzer
d Hufeisen

1405 horseshoe nail
[for]
f clou à ferrer
e clavo de herradura
i chiodo da ferro di cavallo
n beslagnagel; hoefnagel
d Hufnagel

1406 horseshoe pulling pliers
[for/ham]
f tricoise à déferrer
e alicate(s) desmontadoras de herrero
i pinza estrattrice di maniscalco
n hoefijzertang; hoeftang
d Hufbeschlagabreisszange

1407 horsing iron
[nav]
f barrotin; patarasse
e pitarrasa
i paterazzo
n klamaaiijzer
d Klameieisen

1408 hose bit; self-emptying bit
[dril]
f foret à cuiller autoépuratoire
e broca de cuchara autodepuradora
i trapano a cucchiaio autopulitore
n pulslepelboor; zelfreinigende lepelboor
d selbstleerender Löffelbohrer

1409 hose reel
[hor]
f tambour à tuyau
e carrete de la manguera; portamangueras
i tamburo avvolgitubi
n slangehaspel; slangespoel
d Schlauchhaspel

1410 hot chisel
[chi/met]
f tranche à chaud
e tajadera en caliente
i scalpello a caldo
n warmbeitel
d Warmschrottmeissel

1411 hot metal saw
[saw]
f scie pour métaux à chaud
e sierra para metales de régimen caliente
i sega per metalli a caldo
n metaalzaag voor warme objecten
d Metallsäge für heisse Gegenstände

1412 household appliances
[hou]
f outils de ménage
e herramientas de la casa
i arnesi della casa
n huishoudelijk gerei
d Haushaltwerkzeuge

1413 **hydraulic puller**
[pull]
f extracteur hydraulique
e extractor hidráulico
i estrattore idraulico
n hydraulische trekker
d hydraulischer Abzieher

I

1414 ice saw
[hou/saw]
f scie de glace
e sierra de hielo
i sega di ghiaccio
n ijszaag
d Eissäge

* **idler pulley → 1231**

1415 idling pulley
[eng/tra]
f poulie-guide
e polea guía
i puleggia di rimando
n leischijf
d Leitrolle

* **impact crusher → 1259**

1416 impactor forging hammer
[for/ham]
f marteau-pilon à chocs
e martillo pilón de choques
i mazza da fucina ad urti
n met stoten werkende smeedhamer
d mit Stössen wirkender Schmiedehammer

1417 in-cannel gouge; inside gouge
[chi/woo]
f gouge à cambrure interne
e gubia de combeo interno
i sgorbia a curvatura interna
n steekguts met vouw binnen
d Hohlbeitel mit Innenfase

1418 inclinable rotary table
[gen]
f table rotatoire inclinable
e mesa rotatoria inclinable
i tavola rotatoria inclinabile
n schuinstelbare draaitafel
d schrägstellbarer Drehtisch

1419 inclined type of gas pipe tongs
[pli]
f pince à tuyaux de coin
e alicate(s) de tubos esquineros
i giratubi tipo inclinato
n hoekpijpentang
d Eckrohrzange

1420 included angle
[dri/fre/scr]
f angle de noyage
e ángulo de avellana
i angolo di accecamento
n verzinkhoek
d Senkwinkel; Versenkwinkel

1421 increaser; taper pipe
[hyd]
f raccord conique
e cono reductor
i manicotto di riduzione
n verloopstuk
d Übergangsrohr

1422 indented foundation bolt
[fas/iwa]
f boulon de fondation dentelé
e perno de cimentación con indentaciones
i bullone di fondazione indentato
n getande fundatiebout
d gezackter Fundamentbolzen

1423 indexing head
[eng/too]
f poupée à diviser
e cabezal divisor
i testa a dividere
n verdeelkop; verdeeltoestel
d Teilkopf

1424 induction heated teeth
[met/saw]
f denture traitée par trempe à induction
e dentadura caldeada por inducción
i dentatura riscaldata per induzione
n door inductie verhitte vertanding
d induktionsgeheizte Verzahnung

* **industrial diamond dressing file → 728**

1425 inserted blade cutter; inserted tooth cutter
[fre]
f fraise à dents rapportés; fraise à lames rapportées
e fresa de cuchillas postizas
i fresa a lame riportate
n frees met verwisselbare snijtanden
d Fräser mit eingesetzten Schneidzähnen

1426 insertion head
[eng/too]
f tête de chargement
e cabezal de inserción
i testa d'inserzione
n inzetkop
d Einsatzkopf

1427 inside cal(l)ipers
[mea]

f compas d'intérieurs
e compás para interiores
i calibro per interni
n binnenpasser; voetjespasser
d Innentaster; Lochtaster

* **inside gouge → 1417**

1428 **inside micrometer cal(l)ipers**
[meal]
f micromètre d'intérieurs
e micrómetro de interiores
i asta micrometrica per interni
n binnenmicrometer
d Innenmikrometer; Mikrometerstichmass für Innenmessungen

1429 **inspection ga(u)ge**
[meal]
f calibre d'acceptation
e calibre de aceptación; galga de verificación
i calibro di accettazione; calibro di collaudo
n afneemkaliber; revisiekaliber
d Abnahmelehre; Revisionslehre

1430 **inspection lamp**
[con/gen/saf]
f baladeuse
e lámpara portátil
i lampada portatile
n handlamp; looplamp
d Handlampe

1431 **insulated bolt**
[ele/fas/iwa]
f boulon isolé
e perno aislado
i bullone isolato
n geïsoleerde bout
d isolierter Bolzen

1432 **insulated combination pliers**
[pli]
f pince universelle à branches gainées
e alicate(s) universal(es) de mangos plastificados
i pinza universale a manici isolati
n combinatietang met geïsoleerde benen
d Kombizange mit isolierten Schenkeln

1433 **insulated diagonal cutting nippers**
[pli]
f pince coupante diagonale à branches gainées
e alicate(s) de corte diagonal de mangos plastificados
i pinza per taglio diagonale a manici isolati
n diagonaaltang met geïsoleerde benen
d Diagonalschneider mit isolierten Schenkeln

1434 **insulated handles**
[cpt/pli]
f branches gainées
e mangos plastificados
i manici isolati
n geïsoleerde benen
d isolierte Schenkel

1435 **insulated pliers; pliers with insulated handles**
[ele/pli]
f pinces à branches isolées
e alicate(s) con mangos plastificados
i pinze con manici isolati
n tangen met geïsoleerde benen
d Zangen mit isolierten Schenkeln

1436 **insulated shank screwdriver**
[scd]
f tournevis à lame isolée
e destornillador de hoja aislada
i cacciavite a lama isolata
n fittingschroevedraaier; schroevedraaier met geïsoleerde kling
d Schraubendreher mit isolierter Klinge

1437 **insulation stripping pliers; wire stripping pliers**
[pli]
f pince à dénuder
e alicate(s) pelacables
i pinza spellafili
n afstriptang; draadafstriptang; draadstriptang
d Abisolierzange

1438 **integral machines**
[gen]
f intégrales
e máquinas integrales
i macchine integrali
n compacte machines; karwei-klare machines
d Kompaktmaschinen

1439 **interchangeable counterbore**
[dri]
f mèche à tenon interchangeable
e escariador intercambiable
i punta per la battuta finale intercambiabile
n uitwisselbare soevereinboor
d auswechselbarer Zapfenbohrer

1440 **intermediate tap; second tap**
[scr]
f taraud intermédiaire
e macho de roscar segundo; macho intermedio
i maschio intermedio
n tussensnijder
d Mittelschneider

1441 **internal cylindrical ga(u)ge; plug ga(u)ge**
[mea]
f calibre à tampon cylindrique; jauge-tampon
e calibre de agujeros; calibre macho
i calibro a tampone
n gatkaliber; penkaliber
d Lehrdorn

1442 **internal grinder**
[gri]
f rectificatrice d'intérieurs
e rectificadora para interiores
i rettificatrice per interni
n inwendig-slijpmachine
d Innenschleifmaschine

1443 **internal grinding**
[gri]
f rectification intérieure
e rectificado de interiores
i rettifica degli interni
n inwendig slijpen
d Innenschleifen

* **internal screwthread → 961**

* **iron crowbar → 1120**

J

1444 jack
[eng/gen]
f cabre; cric; vérin
e cric; gato
i cric; cricco; martinetto
n dommekracht; krik
d Hebevorrichtung

1445 jackhammer
[ham/min/sto]
f marteau pneumatique à main
e martillo neumático a mano
i martello pneumatico a mano
n pneumatische handhamer
d Drucklufthammer

1446 jack plane
[car/pla/woo]
f riflard
e garlopa
i piallone
n roffel; roffelschaaf; voorloper
d Rauhbank

1447 jacot tool
[wat]
f tour à pivoter; tour Jacot
e torno de pivotear; torno para pulir
i tornietto per lucidare perni
n horlogemakersdraaibank
d Zapfenrollierstuhl

1448 jag bolt; rag bolt
[fas/iwa]
f boulon barbelé
e perno arponado
i bullone a mazzetta
n takbout
d Bartbolzen

* **jag bolt → 1094**

* **jamb brush → 831**

1449 jar
[dri/min]
f coulisse de perforateur
e barra de perforación; corredera
i asta a snodo
n boorstaaf
d Bohrstange

* **jaw → 498**

1450 jaws of pliers
[cpt/pli]
f becs de pinces
e bocas de alicates
i becchi di pinze
n bekken van een tang
d Backen einer Zange

1451 jedding axe
[foy]
f merlin de bûcheron
e hacha de rajar; piqueta
i ascia fenditrice
n kloofbijl
d Spaltaxt

* **jedding axe → 441**

* **jenny cal(l)ipers → 1347**

1452 jeweller's file; needle file; Swiss pattern file
[far]
f lime aiguille
e lima de aguja
i lima ad ago
n naaldvijl; steelvijltje
d Nadelfeile

1453 jeweller's piercing saw; lightweight fret saw; piercing saw
[saw]
f scie de jouaillier
e sierra de joyero
i sega da gioielliere
n instrumentmakerszaag
d Juwelierssäge

1454 jib
[tra]
f flèche de la grue
e aguilón
i braccio della gru
n zwaaiarm van een kraan
d Ausleger

1455 jib barrow
[agr/hor/tra]
f diable
e carretilla sin costadas
i carriola senza sponde
n kruiwagen zonder zijwanden
d Schubkarre(n) ohne Seitenwangen

1456 jig
[gen/lat/mea]
f gabarit de réglage
e plantilla para posicionar
i attrezzatura di guida
n pasmal
d Spannvorrichtung

1457 **jig borer**
[dri]
f foreuse à gabarit
e taladradora de plantillas
i trapano con maschera
n boormachine met mal
d Bohrmaschine mit Schablone

1458 **jig saw**
[saw]
f scie à guichet; scie sauteuse
e miniserrucho; serrucho de punta
i sega a foretto
n schrobzaag
d Stichsäge

1459 **jig saw attachment**
[saw]
f adaptation scie à guichet; adaptation scie sauteuse
e aditamento para miniserrucho; aditamento para serrucho de puntas
i adattamento per sega a foretto
n schrobzaagvoorzetstuk
d Stichsägevorsatz

1460 **jim crow**
[gen/lat]
f porte-outil orientable
e cabezal de herramientas orientable
i corsoio portautensili orientabile
n draaibare gereedschapshouder
d drehbarer Werkzeughalter

1461 **jim-crow**
[tra]
f cintreuse de rails
e dispositivo de curvar carriles
i apparecchio per curvare le rotaie
n railbuiger
d Schienenbiegegerät

1462 **jim-crow**
f pied de chèvre
e palanca de pie de cabra
i piede di porco
n koevoet
d Brechstange mit Nagelklaue

1463 **joggle; stub tenon**
[car/joi/woo]
f tenon invisible
e espiga invisible; espiga oculta
i tenone corto
n verdekte pen
d Fusszapfen

* **joiner's mallet → 415**

* **joiner's mortise chisel → 1104**

1464 **joint**
[car/joi/woo]
f assemblage; chevillage
e ensambladura
i incastro
n verbinding
d Verzapfung

1465 **jointer**
[mas]
f fiche; mirette
e paletín
i ferro per giunti di mattoni
n dag; dagge; dagijzer; voegijzer
d Fugeneisen; Fugenkelle

* **jointer → 335**

1466 **jointer plane; jointing plane; long plane; shooting plane; trying plane; try plane**
[car/pla/woo]
f varlope
e galera
i pialla da sgrossare
n reeschaaf; rijschaaf
d Langhobel; Rauhbank

1467 **joint scraper; pointer**
[mas]
f grattoir du maçon
e rascador para rejuntar
i raschietto di giunti di mattoni
n voegkrabber
d Fugeneisen

K

1468 **kettle; ladle**
[fou/met]
f poche; poêle
e caldera; cuchara
i caldaia; secchia di colata; siviera
n gietpan; smeltketel; smeltpan
d Pfanne; Schmelzpfanne

1469 **kevel; kevel hammer**
[ham/sto]
f casse-pierre(s); marteau à ballast
e martillo de picapedrero
i martello di scalpellino
n steenkloppershamer; steenslaghamer
d Hammer des Steinschlägers; Steinhammer

1470 **keyed chuck**
[dri]
f mandrin à clé
e mandril de llave
i mandrino a chiave
n boorhouder met sleutel
d Zahnkranzbohrfutter

1471 **key file; warding file**
[far]
f lime à clefs
e lima de llaves
i lima per chiavi
n sleutelvijl
d Schlüsselfeile

1472 **keyhole saw; pad saw**
[saw]
f scie à guichet
e serrucho de punta
i gattuccio
n sleutelgatzaag
d Stichsäge

1473 **keyway**
[car/hou]
f trou de clé
e bocallave; chavetero
i buco di chiave
n sleutelgat; spiebaan
d Schlüsselloch; Schlüsselschlitz

1474 **keyway tool**
[car/tool]
f outil de rainurage
e útil de cortar chaveteros
i coltello di stozzatrice per cave; utensile per scanalare
n spiebaanstaal
d Nutstahl

1475 **knapping hammer**
[ham/sto]
f marteau casse-pierres
e almádena; martillo para picar piedra
i martello di scalpellino; mazza di cavatore
n steenbreker
d Keilhammer

1476 **knapping hammer; miner's pneumatic hammer**
[ham/min]
f pic pneumatique de mineur
e martillo picador neumático de minero
i piccone pneumatico di minatore
n pneumatische mijnwerkershamer
d Drucklufthammer für Grubenarbeiter

* **kneading iron → 320**

1477 **kneading table**
[nut]
f table à pétrir
e hintero; mesa de amasar
i tavola per impastare
n bakkersbraak
d Knettisch; Wirktisch

1478 **kneading trough**
[nut]
f pétrin
e amasadera; artesa
i gramola; gramolatrice; madia per impastare
n bakkerstrog
d Backtrog

* **knee → 859**

1479 **knife and scissors sharpener**
[gri/hou/sha]
f affûte-couteaux et ciseaux
e afiladora de cuchillas y tijeras
i affilatrice per coltelli e cesoie
n messen- en scharenslijpapparaat
d Messer- und Scherenschleifgerät

1480 **knife file**
[far]
f lime couteau
e limacuchillo
i lima a coltello
n mesvijl
d Messerfeile

1481 **knife sharpener; sharpening steel; steel**
[gri/hou/nut/sha]

f affiloir de bucher; fusil de boucher
e afilón; chaira
i affilatoio
n aanzetstaal
d Abziehstahl

1482 knife tool
[cpt/lat]
f outil à couper
e herramienta de corte
i utensile a coltello
n snijgereedschap
d Schneidwerkzeug

1483 knife tool *GB*; **offset tool** *US*; **side tool**
[cpt/lat]
f outil couteau
e cuchilla acodillada
i utensile laterale per tornio
n mesbeitel; zijbeitel
d Seitendrehmeissel; Seitenstahl

1484 knuckle
[cpt/fas/iwa]
f charnon
e unión de charnela
i giunto a cerniera
n scharnierpen
d Scharnierstift

1485 knurling tools
[fre/met]
f molettes rayées
e moletillas
i godrone; utensili per zigrinare
n kartelijzer; kartelrol
d Rändelrädchen; Rändelwerkzeug

L

1486 **labyrinth packing**
[eng/saf]
f garniture à labyrinthe
e empaquetadura de labirinto; junta labiríntica
i tenuta a labirinto
n labyrintafsluiting
d Labyrinthdichtung

1487 **ladder**
[fir/hou/saf]
f échelle à barreaux
e escalera de escalones; escalera de mano
i scala a pioli
n sportenladder
d Sprossenleiter

* **ladle → 1468**

1488 **lambswool bonnet**
[cpt/pol]
f bonnet peau de mouton
e funda de piel de cordero
i cuffia lana merinos
n lamsvelhoes
d Lammfellhaube

1489 **laminate cut; laminate cutter**
[car/fre/wool]
f fraise à biseauter les bords en matière artificielle dure
e fresa biseladora para plásticos duros
i fresa per bisellare plastiche dure
n kantfrees voor harde kunststof
d Kantenfräser für harte Kunststoffe

1490 **lance**
[cpt/cut/tool]
f traçoir
e punta precortadora
i punta trinciante
n voorsnijder
d Vorschneider

1491 **lance toothing**
[cpt/saw]
f denture lancéiforme
e dentadura de lanzas
i dentatura a lancie
n speerpuntvertanding
d Lanzenspitzenzahnung

1492 **lap; lapping disk**
[fin/pol]
f disque à rodage
e disco de tribopulimento
i disco da lappatura
n lepschijf
d Läppscheibe

* **lap dovetail → 775**

* **lapidary grinding wheel → 895**

1493 **lap joint**
[car/joi/wool]
f assemblage à recouvrement
e ensambladura a solapa
i giunto a sovrapposizione
n overlappingsverbinding
d Überlappungsverbindung

1494 **lapping**
[cle/gri/pol]
f polissage; rodage à l'abrasif libre
e lapidado; tribopulimento
i lappatura; smerigliatura
n leppen
d Läppen

* **lapping disk → 1492**

1495 **lapping machine**
[fin/pol]
f machine de rodage
e máquina de tribopulimento
i macchina da lappatura
n lepmachine
d Läppmaschine

1496 **larry**
[mas]
f bouloir; broyon à mortier
e azado para hacer la mezcla
i impastatrice di malta
n mortelmenger; roerijzer
d Mörtelmischer

1497 **lateral guide; rip fence**
[cpt/saw]
f guide latéral
e guía lateral
i guida laterale
n langsgeleider
d Längsanschlag

1498 **lateral stake**
[cpt/for]
f fer à rabattre; tranche de ferblanterie
e bigornia lateral
i bicornia laterale
n randstaak
d Umschlageisen

1499 lathe
[cut/lat]
f tour
e torno
i tornio
n draaibank
d Drehbank; Drehmaschine

1500 lathe center(re)s
[cpt/lat]
f pointes de tour
e puntas de torno
i punte di tornio
n draaibankcenters
d Drehmaschinenspitzen

1501 lathe file
[far/met]
f lime de tour; lime pour tourneur
e lima de torno
i lima da tornio
n draaibankvijl
d Drehbankfeile

* **lathe headstock → 1329**

1502 lathe tools; turning tools
[lat]
f outils de tour
e útiles del torno
i utensili di tornio
n draaibankgereedschappen
d Drehbankgeräte

1503 lawnborder shears
[hor]
f coupe-bordures
e tijeras para bordes de césped
i forbici per bordure di prato
n graskantenschaar
d Rasenkantenschere

1504 lawn mower
[hor]
f tondeuse à gazon
e cortacésped; segadora de césped
i tagliaerba; tosatrice da prato
n gazonmaaier
d Rasenmäher

* **lawn mower → 1196**

1505 lawn pricker
[hor]
f aérateur de pelouse
e aerador de césped
i aeratore di prato
n gazonbeluchter; gazonprikker
d Rasenlüfter

1506 lawn rake
[hor]
f râteau de pelouse
e rastrillo de césped
i rastrello di prato
n gazonhark
d Rasenharke

* **lawn shears → 1198**

1507 lawn sprinkler
[hor]
f arrosoir rotatif
e aspersor rotativo del césped
i irroratrice rotativa
n gazonsproeier
d Rasensprenger

* **laying-on trowel → 1004**

* **laying trowel → 339**

1508 lazy tongs
[hou/pli/tra]
f pince à zigzags
e tenazas de resorte
i pinze per prendere oggetti lontani o alti
n Neurenberger schaar
d Nürnberger Schere; Scherenspreizer

1509 lead hammer
[ham/hyd]
f marteau à plomb
e martillo de plomo
i martello per lavorare il piombo
n loodklopper
d Bleihammer

1510 lead nail
[iwa]
f clou de courreur
e clavo de techumbre; clavo para fijar hojas de plomo
i chiodo per lamiere dei tetti
n bladloodspijker; daknagel
d Dachnagel

1511 lead-pipe expander turnpin
[pli]
f embauchoir; pince à élargir; pince à évaser
e ensanchador; mandrinadora
i allargatoio
n loodpijppuitzettang
d Bleirohraufweitezange

1512 leather apron
[for]
f tablier de cuir

e mandil de cuero
i grembiule di cuoio
n schootsvel
d Schurzfell

1513 leather pincers; welt pincers
[lea/pli]
f pince de montage de cordonnier
e alicate(s) para el cuero
i pinza di cuoio
n boordseltang; felstang
d Falzzange

1514 left-hand tools
[gen/lat]
f outils à rotation à gauche
e herramientas de rotación a la izquierda
i attrezzi sinistrorsi
n naar links draaiend draaibankgereedschap
d linksdrehende Drehbankwerkzeuge

1515 lengthening bar
[mea]
f rallonge d'un compas
e alargadera de compás
i prolungamento del gambo
n passerbeenverlenging
d Schenkelverlängerung

1516 lengthening rod
[dri/min]
f barre de rallongement de sonde
e barra de alargamiento de sonda
i sbarra di allungatura di sonda
n verlengstaaf van een boorstang
d Verlängerungsstab einer Bohrstange

1517 lettering brush
[dec]
f pinceau à lettres
e pincel para rótulos
i pennello per i caratteri
n letterpenseel; letterzetter
d Schriftpinsel

1518 level
[mea]
f niveau
e nivel
i livella
n libel; waterpas
d Libelle; Wasserwaage

1519 level(l)ing file
[far/saw]
f lime d'alignement
e lima de alineación
i lima ad allineare
n strijkvijl
d Richtfeile

1520 level(l)ing pole; level(l)ing rod; level(l)ing staff
[mea/sur]
f jalon d'arpentage; mire de nivellement
e mira topográfica; niveleta; vara de agrimensor
i biffa; canna di misura; mira
n bakenstok; meetstok; nivelleerlat
d Absteckpfahl; Nivellierlatte; Nivellierstab

1521 level trier
[mea/sur]
f appareil de mesure de la sensibilité de niveaux
e medidor de la sensibilidad de niveles de burbuja
i apparecchio di prove per livelli
n waterpastestapparaat
d Libellenprüfgerät

1522 level tube; spirit level; vial
[mea]
f fiole de niveau; tube de verre
e tubo del nivel
i tubo di livella
n libel; waterpasbuisje
d Libelle

1523 lever
[gen/tra]
f levier
e palanca
i leva
n hefboom
d Hebel

1524 lever cap; plane cover iron
[cpt/pla]
f bloc d'arrêt
e hierro de bloqueo
i ferro di arresto
n dekbeitel
d Messerklappe

* **lifter** *GB* → **1123**

1525 lifting hook
[foy]
f crochet de levage; sapine
e gancho de alza; gancho de izar
i gancio di sollevamento; leva ad uncino
n boomhevel
d Sapine; Sappel

1526 lifting screws
[fou]

f vis de soulèvement
e gatos de tornillo; husillos de izado
i viti di sollevamento
n hefschroeven
d Hebeschrauben; Hubschnecken

1527 **lifting shears**
[civ/tra]
f cisaille à lever
e cizalla de palanca
i cesoia a leva
n heftang
d Hebelschere

* **lightweight fret saw → 1453**

1528 **limit ga(u)ge**
[mea]
f calibre de limites; calibre de tolérance
e calibre de límites; calibre de tolerancia
i calibro differenziale; calibro di tolleranza
n grenskaliber
d Grenzlehre

1529 **limpet washer**
[fas/gen]
f rondelle de fixation
e arandela de fijación
i rondella di fissazione
n bevestigingsringetje
d Befestigungsscheibe

1530 **linch pin**
[saf/tra]
f esse
e pezonera; sotrozo
i spinotto
n luns
d Radzapfen

1531 **line drawing pencil; liner**
[dec]
f pinceau traîneur
e pincel trazador
i pennello transcinatore
n biezentrekker; sleper
d Schlepper

1532 **line pins**
[mas]
f chevilles de cordeau
e pernos de cuerda
i perni di corda
n metselkoordpennen
d Schnurpinnen

* **liner → 1531**

1533 **linesmen's pliers; telephone pliers**
[pli]
f pince pour ouvriers téléphoniques
e tenaza de montador telefonista
i pinza per telefonisti
n telefoontang
d Telephonzange; Zange für Fernsprech-techniker

1534 **live center(re); running center(re)**
[lat]
f pointe tournante
e contrapunta giratoria
i punta della controtesta mobile; punta della fontina
n meelopende spilpunt
d mitlaufende Reitstockspitze; umlaufende Reitstockspitze

1535 **loader**
[min]
f pelle mécanique
e pala mecánica
i pala meccanica
n mechanische laadschop
d maschinelle Schaufel

1536 **lock**
[hou/saf]
f serrure
e cerradura
i serratura
n slot
d Schloss

* **lock bolt chisel → 278**

* **locking key → 1081**

1537 **lock mortise chisel; swan neck mortise chisel**
[chi]
f bédane cuillère
e cincel de cuchara
i bedano in forma di cucchiaio
n lepelschietbeitel; lepelvormige schietbeitel
d löffelförmiger Lochbeitel; Löffellochbeitel

1538 **lock nut**
[fas/iwa]
f contre-écrou
e contratuerca
i controdado
n contramoer
d Gegenmutter; Kontermutter

* **lock nut → 477**

* **lock saw → 576**

* **locksmith's chisel → 648**

1539 **locksmith's hammer**
[ham]
f marteau de serrurier
e martillo de cerrajero
i martello di fabbro ferraio
n slotenmakershamer
d Schlosserhammer

1540 **locksmith's rasp**
[far/met]
f râpe serrurier
e escofina de cerrajero
i raspa di chiavaiuolo; raspa di magnano
n slotenmakersrasp
d Schlosserraspel

1541 **lock spring**
[cpt/iwa/saf]
f ressort d'arrêt; ressort de blocage; ressort de sûreté
e muelle enclavador
i molla di bloccaggio; molla di blocco
n borgveer
d Sicherungsfeder

1542 **logging wheels; timber bob; timber janker**
[foy]
f trique-balle
e carro remolque; diablo para transporte de troncos; zorrilla portatroncos
i baroccio; bilanciere; ciondolo
n mallejan
d Holzrückwagen; Rückewagen

1543 **log saw; pit saw**
[foy/saw]
f scie à bûches; scie à tronçonner; scie forestière
e sierra arbolera; sierra de tronzar; sierra forestal
i sega per tronchi
n afkortzaag; boomzaag; kortzaag
d Ablängsäge; Baumsäge; Waldsäge

1544 **long chain nose pliers; radio pliers**
[pli]
f pince à becs longs aplatis
e alicate(s) de bocas planas; largas y picos rectos
i pinza a becchi piatti lunghi
n tang met lange platte bekken
d Langbeckzange mit flachen Backen

* **long-handled pruner → 109**

1545 **longitudinal joint**
[car/joi/woo]
f assemblage en long
e ensambladura longitudinal
i incastro longitudinale
n lengteverbinding
d Längenverbindung

* **long plane → 1466**

1546 **long shank box spanner; nut runner**
[scr]
f clé de serrage d'écrous
e apretatuercas mecánico
i chiave per avvitare
n moeraanzetter
d Mutteranzieher

1547 **long snipe nose pliers**
[pli]
f pince à becs longs affilés
e alicate(s) de bocas planas largas y puntiagudas
i pinza a becchi piatti lunghi appuntati
n tang met lange platte spits uitlopende bekken
d Langbeckzange mit spitz zulaufenden flachen Backen

1548 **loose pulley**
[cpt/tra]
f poulie folle
e polea loca
i puleggia a folle
n losse katrol
d Leerlaufrolle

* **lopping billhook → 239**

1549 **lopping chisel**
[agr/hor/foy]
f ciseau à ébrancher; ciseau à élaguer
e cincel desramador; cincel para podar
i scalpello per diramare
n boombeitel; snoeibeitel
d Astschneider; Putzbeitel

1550 **low-angle plane**
[pla]
f angle de coupe de 12°
e cepillo de ángulo de corte 12°
i pialla con inclinazione del ferro a 12°
n beitel met snijhoek van 12°
d Beitel mit 12°-Schnittwinkel

1551 **lower case**
[pri]
f bas de casse
e caja baja

i cassa di sotto; minuscolo
n onderkast
d Unterkasten

* **lowering wedges → 839**

* **lump hammer → 540**

M

* **machete** → 373

1552 **machine countersink**
[dri/real]
f alésoir-fraise mécanique
e avellanador mecánico
i accecatoio meccanico
n machineverzinkfrees
d Maschinensenker

1553 **machine reamer**
[dri/fre/real]
f alésoir mécanique
e escariador mecánico
i alesatore meccanico
n machineruimer
d Maschinenreibahle

1554 **machine tools**
[gen/tool]
f outils de machines
e herramientas de máquinas
i arnesi di macchine; utensileria di macchine
n machinegereedschappen
d Maschinenwerkzeuge

1555 **machine tools**
[gen/tool]
f machine-outils
e máquina-herramientas
i macchina-utensili
n gereedschapsmachine
d Werkzeugmaschine

* **machine vice** → **1049**

1556 **magnesite-bonded abrasive**
[gri/san]
f abrasif à agglomérant magnésite
e abrasivo ligado con magnesita
i abrasivo ad impasto di magnesite; abrasivo collato con magnesite
n magnesiet-gebonden slijpmiddel
d magnesitgebundenes Schleifmittel

1557 **magnetic chuck**
[gri/tool]
f mandrin magnétique
e mandril magnético
i mandrino magnetico
n elektromagnetische boorhouder
d elektromagnetisches Futter; Magnetspannfutter

* **magnetic compass** → 575

1558 **magnetic drill-stand**
[dri]
f support magnétique
e bastidor magnético
i supporto verticale magnetico
n magnetische machineboorstandaard
d Magnetbohrständer

1559 **magnetic separator**
[cle/sep]
f séparateur magnétique
e separador magnético
i separatore magnetico
n magnetische scheider
d magnetische Trenn- und Sortiervorrichtung; Magnetscheider

1560 **mahlstick; maulstick**
[art/dec]
f appui-main
e apoyomano; tiento
i bacchetta canna; mazza di pittore
n schildersstokje
d Malerstock

1561 **mains tester; voltage detector; voltage indicator**
[con/ele]
f déceleur de tension; détecteur de tension
e detector de tensión; indicador de tensión
i indicatore di tensione
n spanningsverklikker; spanningzoeker
d Spannungsanzeiger; Spannungsprüfer

1562 **mallet**
[ham]
f marteau
e martillo grande de madera; mazo
i mazzuola
n houten hamer
d Holzhammer

1563 **mandrel; mandril**
[cpt/tool]
f mandrin
e mandril
i mandrino
n gereedschapshouder
d Futter; Spannfutter; Werkzeughalter

1564 **manufacturing ga(u)ge; working ga(u)ge**
[mea]
f calibre d'atelier; calibre de fabrication
e calibre de fabricación; calibre de taller
i calibro di fabbricazione; calibro d'officina
n fabricagekaliber; werkplaatskaliber
d Fertigungslehre; Werkstattlehre

1565 manure spreader
[agr]
f épandeur de fumier
e distribuidor de abonos
i spandiletame
n meststrooier
d Dungstreuer; Kreiseldüngerstreuer

1566 map measurer
[mea/sur]
f curvimètre; molette métrique
e curvímetro; opisómetro
i curvimetro
n curvimeter van landkaarten; meetraadje
d Messrad für Landkarten

1567 marble saw
[saw]
f scie à marbre
e sierra para mármol
i sega per marmo
n marmerzaag
d Marmorsäge

1568 marbling brush
[dec]
f pinceau à marbrer
e pincel jaspeador
i pennello per marmorizzare
n marmerkwast
d Marmorierpinsel

* **marbling comb → 560**

1569 margin trowel; plasterer's trowel
[mas]
f truelle de plâtrier
e paleta de estucador
i frettazzo di stuccatore
n bandtroffel; stukadoorstroffel
d Stukkateurkelle

1570 marker ***US*****; marking hammer** ***GB*****; marking iron; revolving die hammer**
[foy/ham]
f marteau numéroteur rotatif
e numerador giratorio de revolver
i martello numeratore giratorio
n ritshamer; roterende merkhamer; roterend merkijzer
d Revolvernumerierschlegel

1571 marking ga(u)ge; mortise ga(u)ge
[mea]
f tracequin; trusquin
e gramil de orilla
i graffietto
n kruishout
d Streichmass

* **marking hammer** ***GB*** **→ 1570**

* **marking iron → 1570**

1572 marking knife
[mea]
f couteau à tracer
e cuchilla trazadora
i coltello a tracciare
n afschrijfmes
d Reissmesser

1573 marline spike
[fas/nav]
f épissoir; poinçon à épisser
e pasador de empalmar
i caviglia da allacciare insieme
n gatnaald; marlpriem; marlspijker
d Marlpfriem; Marlspicker; Splisseisen

1574 masonry drill
[dri]
f foret bâtiment; trépan bâtiment
e barrena para piedras
i punta elicoidale per pietre, cemento e mattoni
n steen- en betonboor
d Mauerbohrer; Steinbohrer

1575 masonry saw; stonemason's saw
[saw]
f scie à substances pierreuses
e sierra para cortar sillares
i sega per sostanze pietrose
n tweemanssteenzaag
d Allzwecksteinsäge

1576 mason's bolster; stone chisel
[chi/sto]
f ciseau de tailleur
e cincel de picapedrero
i scalpello di tagliapietre
n ceseel; slagbeitel
d Breiteisen; Steinmetzmeissel

* **mason's bolster → 336**

1577 mason's carving chisels and gouges
[chi/sto]
f ciseaux graveurs de caractères
e cinceles grabadores de caracteres
i scalpelli intagliatori di caratteri
n letterbeitels
d Buchstabenmeissel

1578 mason's chisel
[chi/sto]
f ciseau lourd de tailleur
e cincel pesado de picapedrero

i scalpello pesante di tagliapietre
n job; jop
d schwerer Steinhauermeissel

1579 **mason's level; plummet level**
[mas/mea]
f niveau de maçon
e nivel de albañil
i livella di muratore
n metselaarswaterpas
d Maurerwasserwaage

* **mason's level → 1101**

1580 **mason's rule**
[mas/mea]
f mètre de maçon
e metro de albañil
i metro di muratore
n metselaarsduimstok
d Maurermesstab

1581 **master gauge** *GB*; **reference gage** *US*
[mea]
f rapporteur
e calibre de referencia; galga patrón
i calibro campione; calibro di risconto
n standaardkaliber
d Kontrollehre; Prüflehre; Vergleichlehre

1582 **master tap**
[cpt/dri/scr]
f taraud finisseur
e macho matrón de roscar
i maschio di rifinitura
n afwerktap; fijntap
d Feinschneider

* **mastic gun → 435**

* **mastic trowel → 14**

1583 **matched boards**
[car/joi/woo]
f planches bouvetées
e tablas machihembradas
i assi appaiati a maschio e femmina
n geploegde planken
d Riemenbretter

* **matchet → 373**

1584 **matching planes; match planes**
[pla]
f jeu de bouvets à joindre; jeu de bouvets mâle et femelle
e grupo de cepillos de machihembrar
i gruppo di pialle per linguetti e scanalature
n stel veer- en messingschaven
d Satz von Feder- und Nuthobeln

1585 **matrix holder**
[gen/too]
f porte-matrice; support
e portamatriz
i portastampo
n matrijshouder
d Matrizenhalter

* **matt(e) grinding → 826**

1586 **mattock**
[agr/car/gen]
f houe; pioche
e zapapico
i gravina
n piekhouweel (punt en beitel); pikhouweel
d Karst; Rodehacke

* **maul → 187**

* **maulstick → 1232, 1560**

1587 **measuring tapes**
[gen/mea]
f mesures à ruban; mètres à ruban; rubans métriques
e cintas métricas
i flessometri; metri a nastro; nastri metrici
n bandmaten; maatbanden; meetlinten
d Bandmasse; Massbänder

1588 **measuring wheel; odometer; perambulator**
[mea/sur]
f podomètre
e podómetro
i podometro
n meetwiel
d Messrad; Wegrad

1589 **melting ladle**
[hyd/met]
f cuiller de plombier
e cazo de plomero
i cucchiaio di idraulico
n loodlepel
d Bleilöffel

1590 **melting pot; crucible**
[met]
f creuset
e crisol
i crogiolo
n smeltkroes
d Schmelztiegel

1591 **metal bending machine**
[ben/met]
f machine à plier des métaux
e plegadora para metales
i piegatrice per metalli
n metaalbuigmachine
d Metallbiegemaschine

1592 **metal cutting saw blade**
[cpt/saw]
f lame de scie à métaux
e hoja de sierra para metales
i lama di sega per metalli
n metaalzaagblad
d Metallsägeblatt

* **metal float → 1004**

1593 **metallic packing**
[civ/fas/saf]
f garniture métallique
e empaquetadura metálica
i guarnizione metallica
n metaalpakking
d Metallpackung

1594 **metal miter(re) box**
[saw]
f machine à onglets
e ingletadora profesional
i tagliacornici professionale
n verstekzaagmachine
d Präzisionsgehrungslade

1595 **metal saw**
[saw]
f scie à métaux
e sierra de metales
i sega per metalli
n metaalzaag
d Metallsäge

1596 **metal slitting saw**
[saw]
f scie circulaire pour couper métaux
e sierra circular para cortar metales
i sega circolare per tagliare metalli
n metaalcirkelzaag
d Metallkreissäge

* **metal strap → 357**

1597 **middle gouge**
[car/chi]
f gouge à cambrure moyenne
e gubia de combeo medio
i sgorbia a curvatura media
n guts met gemiddelde welving
d Hohlbeitel mit mittlerer Wölbung

1598 **mill**
[gen/sto]
f broyeur; moulin
e molino
i mulino
n molen
d Mühle

1599 **mill**
[gen]
f usine
e fábrica
i fabbrica
n fabriek
d Fabrik

1600 **milled**
[met]
f crènelé; entaillé
e con cordón
i intraccato; zigrinato
n gekarteld
d gerändelt

1601 **milled**
[met]
f godronné
e moleteado
i godronato
n gegleufd
d geschlitzt

* **milled file → 680**

1602 **milled file blade**
[far]
f plaque fraiseuse
e plato fresador
i placca fresatrice
n freesvijlblad
d Fräserfeilenblatt

1603 **milled file frame; milled file head**
[cpt/far/fre]
f monture pour plaques fraiseuses
e soporte para lima-fresa
i supporto per lima-fresa
n freesvijlbladhouder
d Fräserfeilenblatthalter

1604 **milled file with curved teeth**
[far/fre]
f plaque fraiseuse à denture curviligne
e lima-fresa con dentadura curvilínea
i lima-fresa con dentatura curvilinea
n freesvijlblad met gebogen vertanding;
freesvijlblad met radiuskap
d bogenverzähntes Fräserfeilenblatt

1605 **milled file with straight teeth**
[far/fre]
f plaque fraiseuse à denture oblique
e lima-fresa con dentadura oblicua
i lima-fresa con dentatura obliqua
n freesvijlblad met schuin gefreesde tanden
d schrägverzähntes Fräserfeilenblatt

1606 **milled-head screw**
[iwa/scr]
f vis à tête godronnée; vis à tête moletée
e tornillo de cabeza moleteada
i vite a testa godronata
n schroef met gleufkop
d Schlitzkopfschraube

* **miller arbor → 81**

1607 **mill file; mill saw file**
[far]
f lime plate à scies
e lima plana para sierras
i lima piatta per seghe
n cirkelzaagvijl; molenzaagvijl; raamzaagvijl
d Mühlsägefeile

1608 **milling**
[fre]
f fraisage
e fresado
i fresatura
n frezen
d Fräsen

* **milling arbor → 679**

1609 **milling cutter**
[fre]
f fraise
e fresa
i fresa
n frees
d Fräser

* **milling cutter → 678**

1610 **milling machine**
[fre]
f fraiseuse; machine à fraiser
e fresadora
i fresatrice
n freesmachine
d Fräsmaschine

* **milling machine → 692**

1611 **milling plane**
[fre/pla]
f rabot d'enlèvement à la fraise
e cepillo desfresador
i fresatrice a pialla
n affreesschaaf
d Abfräshobel

* **milling table → 698**

* **mill saw file → 1607**

1612 **millstone dresser**
[cle/sto]
f marteau à meules
e pico para piedra de amolar
i equilibratore di macchina da mulino
n bil; bilhamer; bilijzer
d Mühlpicke

1613 **miner's hammer; pick-axe**
[ham/min]
f marteau d'abattage
e martillo picador de minero
i martello di minatore; piccone di minatore
n afbouwhamer; pikhamer
d Abbauhammer

* **miner's lamp → 708**

* **miner's pneumatic hammer → 1476**

1614 **miner's saw**
[min/saw]
f scie de mineur
e sierra de minero
i sega da minatore
n mijnwerkerszaag
d Bergmannsäge

1615 **miser**
[dri]
f tarrière à gravier
e trépano grande para abrir agujeros en terrenos húmedos
i succhiello per terreno paludoso
n moerasgrondboor
d Sumpfbodenbohrer

1616 **miter(re); miter(re) joint; mit(e)ring**
[joi/woo]
f assemblage à onglets
e ensambladura a hebra; junta a inglete
i giunto ad angolo; giunto ad ugnatura
n verstekverbinding
d auf Gehrung verbunden

1617 **miter(re) block**
[car/saw/woo]
f boîte à onglets
e caja de ingletes

i tagliacornici
n verstekblok
d Gehrungsblock

1618 miter(re) box
[cpt/saw]
f boîte à onglets
e ingletadora
i cassetta per segate cornici
n verstekbak; versteklade
d Gehrungslade

1619 miter(re) clamp; miter(re) cramp
[car]
f presse à onglet
e mordazas de ingletes; sargento para ingletes
i morsetto angolare
n verstekklem
d Gehrungsspanner; Gehrungszwinge

1620 miter(re) dovetail; secret dovetail
[car/joi/wool]
f queue d'aronde cachée
e cola de milano oculta
i giunto ad angolo
n verborgen zwaluwstaart
d verdeckter Zinken

* **miter(re) dovetail → 764**

1621 miter(re) fence
[saw]
f guide à onglets
e guía de inglete
i guida ad ugnatura
n verstekgeleider
d Gehrungsanschlag

* **mit(e)ring → 1616**

* **miter(re) joint → 1616**

* **miter(re) saw → 123**

1622 miter(re) saw-cut; miter(re) sawing board
[saw]
f scie à onglets
e sierra para cortar al sesgo
i sega per cornici
n verstekzaagmachine
d Gehrungssäge

1623 miter(re) square
[car/meal]
f équerre à onglet
e cartabón de inglete; escuadra de inglete; traza ingletes
i squadra ad ugnatura; squadra a sghembo; squadra zoppa
n verstekhaak
d festes Gehrmass; Gehrdreieck

1624 mole grip; plier wrench; self-grip wrench; vice-grip wrench
[pli]
f pinces-étau "vise-grip"
e mordazas de amovire regulables "poligrip"
i pinze articolate "gripi"
n snelklemtang; vadermoorder; vise-grip
d Grip-Zange

1625 mole-trap
[agr/hor]
f piège à taupes; taupière
e trampa para topos
i trappola per talpe
n molleklem
d Maulwurffalle

1626 monkey spanner; monkey wrench
[spa]
f clé à molette; clé anglaise
e llave inglesa
i chiave inglese
n Engelse sleutel
d Engländer

1627 mortice; mortise
[car/wool]
f mortaise
e mortaja
i mortasa
n gat voor een sluitstuk
d Aussparung; Stemmloch; Zapfenloch

1628 mortise and tenon joint; tenon and mortise joint
[car/joi/wool]
f assemblage à mortaise et tenon
e ensambladura de mortaja y muñón
i incastro a mortasa e tenone
n pen- en gatverbinding
d Verzapfung

* **mortise chisel → 1104**

1629 mortise ga(u)ge
[car/meal]
f trusquin à deux pointes
e gramil de doble punta; gramil de mortaja
i graffietto a due punte
n dubbel kruishout; penkruishout
d Doppelstreichmass

* **mortise ga(u)ge → 634, 1571**

1630 motor cradle
[cpt/gen]
f berceau de moteur
e suspensión de motor
i supporto di motore
n motorhouder
d Motorhalter

1631 motor unit
[cpt/gen]
f bloc-moteur
e grupo motor; motor de accionamiento
i motore di azionamento
n motorblok
d Antriebsmotor

*** mo(u)ldboard → 330**

*** mo(u)lding box → 1030**

1632 mo(u)lding cutter
[fre]
f fraise à profiler; fraise de profilage
e fresa de perfiles
i fresa per profilare
n profielfrees
d Profilfräser

1633 mo(u)lding plane
[pla]
f rabot à profiler
e cepillo de perfiles
i sponderuola ad intaccature
n profielschaaf
d Profilhobel

1634 mounting arbor
[cpt/rea]
f broche porte-alésoir creux
e sujetidor postizo
i portaalesatore cavo
n pennentang; ruimerboor; ruimerboorspil
d Aufsteckhalter

1635 mouth of a plane; slit of a plane
[pla]
f lumière d'un rabot
e abertura de cepillo
i apertura di pialla
n bek van een schaaf
d Spanloch eines Hobels

1636 moving blade
[cpt/cut]
f couteau mobile
e cuchillo móvil
i coltello mobile
n bewegend mes
d Stösselmesser

*** multiple drilling machine → 1124**

*** multiple saw → 1129**

1637 multiple-shape sharpening stone
[gri/sha]
f pierre à affûter multiforme
e piedra de amolar multiforma
i pietra da affilare multiforme
n veelvormige aanzetsteen
d Multiformschleifstein

*** multiple thread → 823**

1638 multiple threaded screw; multi-start thread
[scr]
f vis à plusieurs pas
e tornillo de múltiple rosca
i vite a più filetti
n schroef met meer windingen
d mehrgängige Schraube

1639 multi-purpose plane
[pla]
f rabot universel
e cepillo universal
i pialla universale
n universele schaaf
d Allzweckhobel

*** multi-start thread → 1638**

N

* **nail gun → 421**

1640 nail puller
[car/pli]
f tire-clous
e arrancaclavos; desclavador
i levachiodi
n pompspijkertrekker
d Kistenöffner mit Stahlzange; Nagelzieher

1641 nail punch; nail set
[car/iwa]
f chasse-clous
e botador; sacaclavos
i cacciachiodi
n doorslag; drevel
d Durchschläger; Durchtreiber

1642 natural gas blow lamp
[wel]
f chalumeau à souder à gaz naturel
e lámpara de soldar de gas natural
i lampada per saldare a gas naturale
n aardgassoldeerbrander
d Erdgaslötbrenner

1643 nautical log
[mea/nav]
f loch
e corredera
i solcometro
n log
d Log

* **nebulizer → 94**

* **neck of a chisel → 492**

* **needle file → 1452**

1644 needle roller bearing
[eng/tra]
f palier à aiguilles
e cojinete de rodillos en aguja
i cuscinetto a rullini
n naaldlager
d Nadellager

1645 neoprene-foam coated platen
[cpt/san]
f plateau en mousse néoprène
e platillo lijador neoprenado
i piatto levigatore ricoperto con neoprene
n met neopreenschuim bekleed schuurblad
d Schleifteller mit Neoprenschaum-bekleidung

* **nest saw → 1129**

1646 nib
[met/woo]
f pointe de pied de chèvre
e pico de palanca
i punta di palanchino
n punt van een koevoet
d Brechstangenspitze

1647 nibbler
[cut]
f grignoteuse
e recortadora con punzonado rápido
i spinatrice
n knabbelapparaat
d Knabber

1648 nibbler die
[cut]
f matrice de grignoteuse
e matriz de recortadora
i matrice di spinatrice
n matrijs van het knabbelapparaat
d Knabbermatrize

1649 nibbling
[cut/met]
f découpage à vive arête; grignotage
e recorte de chapa con punzonado rápido
i rosicchiamento; spinatura
n knabbelen
d Knabbern; Knabberschneiden; Nibbeln

1650 nibbling cutter
[cut/met]
f grignoteuse
e recortadora de chapa
i spinatrice
n knabbelschaar
d Knabberschere

1651 nicker; spur
[cpt/dri]
f pointe à couper
e punta para cortar
i punta per tagliare
n snijkantpunt; voorsnijder
d Schneidespitze

1652 nippers; tongs
[car/pli]
f tenaille(s)
e tenaza(s)
i tenaglie
n knijptang; nijptang
d Beisszange; Kneifzange

1653 nippers; stone tongs
[pli]
f tenaille(s) pour soulever des grandes pierres
e tenazas para levantar piedras pesadas
i tenaglie per il sollevamento dei sassi
n steenschaar; steentang
d Steinträgerzange

1654 nipple
[fas/tra]
f écrou d'un rayon de roue
e manguito roscado
i raccordo a vite di raggio
n nippel
d Nippel

1655 nipple
[gen]
f jonction filetée; raccord fileté
e boquilla roscada; manguito roscado
i raccordo interno a vite; raccordo interno filettato
n verbindingsbuis met schroefdraad
d Fitting; Rohrverbindungsstück

1656 nose bit
[dri]
f taraud-cuillère à pointe resserrée
e barrena de cuchara con eliminador de virutas
i trapano a cucchiaio con eliminatore di trucioli
n lepelboor met spaanruimer
d Löffelbohrer mit Spanbeseitiger

1657 nose key
[car/joi/woo]
f contre-clavette
e contrachaveta; llave de pico
i controchiavetta
n tegenspie
d Nasenkeil

* **nose of a plane → 1285**

1658 notched stake
[cpt/for]
f suage à table bombée crénelée
e bigornia de hojalatero
i incudinetto di lattoniere
n vorenstaak
d Sickenstock

1659 nozzle reamer
[bur/fre/rea]
f alésoir pour gicleurs
e escariador de toberas
i alesatore di imboccature
n verstuiverruimer
d Düsenreibahle

* **number sizes → 813**

1660 nut
[gen/iwa]
f écrou
e tuerca
i dado
n moer
d Mutter

1661 nutcracker
[hou/nut]
f cassenoix
e cascanueces
i schiaccianoci
n notenkraker
d Nussknacker

1662 nut pliers
[pli]
f pince à écrous
e alicate(s) para tuercas
i pinza per dadi
n moertang
d Mutterzange

* **nut runner → 1397, 1546**

1663 nylon hammer
[ham]
f marteau à embout nylon
e martillo de cabeza nylon
i martello a testa in nailon
n nylonhamer
d Nylonhammer

O

1664 **oakum comb**
[tex]
f peigne d'étoupe de chanvre
e peine de estopa
i pettine di canapa; trapanata
n hedekam; werkkam
d Wergkamme

1665 **object staff; offset rod**
[mea/sur]
f jalon
e jalón
i stadia
n bakenstok; jalon
d Distanzlatte; Zielstange

1666 **oblique halved joint**
[joi/woo]
f assemblage oblique à mi-bois
e empalme oblicuo a media madera
i giuntura obliqua a mezzo legno
n schuine liplas
d schräge Halbholzverbindung

1667 **oblique scarf joint**
[car/joi/woo]
f assemblage oblique à trait de Jupiter simple
e empalme a rayo de Jupiter
i calettatura a dardo di Giove; giunto a mezzo e mezzo con piano obliquo a dentello
n schuine haaklas
d schräges Hakenblatt

* **oblong punch → 647**

1668 **octogonal set square; octogonal square**
[mea]
f équerre octogonale
e escuadra octagonal
i squadra ottagonale
n achtkantwinkelhaak
d Achteckwinkel

* **odd leg cal(l)ipers → 1347**

* **odometer → 1588**

* **offset rod → 1665**

1669 **offset saw blade**
[cpt/saw]
f lame déportée
e hoja de sierra desplazada
i lama di sega spostata
n verzet zaagblad
d versetztes Sägeblatt

1670 **offset scale**
[sur]
f échelle de perpendiculaires
e escala de ordenadas
i scala di ordinate
n ordinatenschaal
d Ordinatenskala

* **offset spanner → 598**

* **offset tool** *US* **→ 1483**

1671 **ogee bit; ogee cutter**
[car/fre/woo]
f fraise pour doucine
e fresa para ojivas
i fresa per ogive
n ojieffrees
d Spitzbogenfräser

1672 **ogee plane**
[car/pla]
f doucine
e cepillo de cima riversa
i pialla per ogiva
n ojiefschaaf
d Rinnleistenhobel

1673 **oil can**
[eng/lub]
f burette à huile
e aceitera
i oliatore
n oliekan
d Ölflasche; Ölkännchen; Ölkanne

1674 **oiling ring**
[eng/lub]
f anneau de graissage
e anillo de engrase
i anello di lubricazione
n smeerring
d Schmierring

1675 **oil pump**
[eng/lub]
f pompe à huile
e bomba de aceite
i pompa dell'olio
n oliepomp
d Ölpumpe

1676 **oil stone**
[sha]
f pierre naturelle à l'huile
e piedra de aceite

i pietra oleosa
n oliesteen
d Ölstein

* **oilstone slip → 1184**

1677 old woman's tooth; router plane
[pla]
f rabot à corniches; rabot debout; rabot gorget
e guimbarda
i ferro per tracciare
n grondschaaf; horletoet; hoerekind
d Grundhobel; Treppennuthobel

1678 one-man cross-cut saw
[foy/saw]
f scie forestière à une main
e sierra forestal para un solo hombre
i sega diritta ad uno mano
n eenmanstrekzaag
d Einhandstossäge

* **one-man cross-cut saw → 356**

1679 open-ended spanner
[spa]
f clé double fourche
e llave de dos bocas
i chiave fissa doppia
n dubbele steeksleutel
d doppelter Gabelschlüssel; doppelter Maulschlüssel

1680 oscillating garden sprinkler
[hor]
f arroseur oscillant
e aspersor oscilante
i irroratore oscillante
n zwenksproeier
d Schwenksprenger

1681 out-cannel gouge; outside gouge
[car/chi]
f gouge à cambrure externe
e gubia de combeo externo
i sgorbia a curvatura esterna
n vermetguts met vouw buiten
d Hohlbeitel mit Aussenfase

* **outer compasses → 305, 395**

* **outside cal(l)ipers → 305, 921**

1682 outside micrometer cal(l)iper
[mea]
f micromètre d'extérieurs
e micrómetro de exteriores
i micrometro d'esteriori
n buitenmicrometer
d Aussenmikrometer

1683 oval brush
[dec]
f broche ovale
e pincel oval
i pennello ovalo
n ovale kwast
d Ovalpinsel

* **oval file → 983**

1684 overlapping joint
[car/joi/woo]
f assemblage à recouvrement
e junta de recubrimiento; junta de solape
i giunto a sovrapposizione
n potdekselen
d Überlappungsverbindung

P

1685 pad
[cpt/gen/tool]
f poignée universelle
e empuñadura universal
i impugnatura universale
n universele handgreep
d Allzweckhandgriff

1686 padlock
[iwa/saf]
f cadenas
e candado
i lucchetto
n hangslot
d Vorhängeschloss

* **pad saw → 1472**

* **pail → 355**

1687 paint brush; painter's brush
[dec]
f brosse à peindre
e brocha de pintura; pincel
i pennello di colore
n verfkwast
d Malerpinsel

1688 paint burning lamp; painter's torch
[dec]
f chalumeau pour brûlage des peintures
e lámpara para quemar la pintura; lamparilla para ablandar la pintura vieja
i lampada per bruciare la pittura vecchia
n verfafbrander
d Farbabbrennapparat

* **painter's brush → 1687**

1689 painter's shave hook; shave hook
[dec]
f grattoir de peintre
e rasqueta triangular de pintor
i raschietto per decoratore
n verfkrabber
d Farbkratzer

* **painter's torch → 1688**

1690 paint gun; paint sprayer
[dec]
f pistolet à peinture; pulvérisateur pour peinture
e pistola para pintar; pistola rociadora de pintura
i pistola per verniciatura a spruzzo
n verfspuit
d Farbspritzpistole

1691 paint-gun mask
[dec]
f masque pur-air
e estarcido para pintar con pistola
i maschera con pistola a spruzzo
n verfspuitmasker
d Farbspritzermaske

1692 paint mixer
[dec]
f mélangeur de peinture
e mezclador de pintura
i mescolatrice di colore
n verfmenger
d Farbmixer

1693 paint pot
[dec]
f camion de peintre
e tarro de pintura
i recipiente di colore
n verfpot
d Farbtopf

1694 paint roller
[dec]
f broche sans fin; rouleau de peinture
e pintador; rulo de color; rulo de pintura
i rullo di colore
n verfrol; verfroller
d Farbroller

1695 paint scraper; stripping knife
[dec]
f couteau-racloir
e cuchilla de raspar
i coltello per la rimozione della vernice vecchia
n afsteekmes
d Farbabstreifmesser

* **paint spray → 31**

* **paint sprayer → 1690**

1696 paint stirrer
[dec]
f agitateur de peinture; hélice agitatrice
e agitador de pintura
i agitatore di colore
n verfroerder
d Farbenrührer

1697 paint tin
[dec]
f bidon de couleur; boîte de peinture

e lata de pintura
i scatola di colore
n verfbus
d Farbbüchse

* **pairing knife → 1339**

1698 palette
[art/dec]
f palette
e paleta
i tavolozza
n palet
d Palette

1699 palette knife
[art/dec]
f amassette
e espátula de pintor
i spatola di pittore
n paletmes; tempermes; verfmes; verfspatel
d Malerspachtel

1700 panel ga(u)ge
[car/mea]
f trusquin pour surfaces larges
e galga para superficies anchas
i graffietto per superfici larghi
n kruishout voor brede oppervlakken
d Streichmass für breite Oberflächen

1701 panel saw
[car/saw]
f scie à coupe lisse
e surrucho de mano de dientes finos
i sega di mano a denti fini
n fijne handzaag
d Feinsäge

1702 panhead rivet
[iwa/riv]
f rivet à tête tronconique
e remache de cabeza troncocónica
i ribattino con testa a bottone
n klinknagel met afgeplatte-kegelkop
d Kegelsenkkopfniete

1703 panic bolt
[iwa/saf]
f crémone de sûreté
e falleba de emergencia; falleba de salida
i chiavistello di sicurezza
n panieksluiting
d Panikverschluss

1704 paper disk
[cpt/gri/san]
f disque à papier abrasif
e disco de papel abrasivo
i disco a carta abrasiva
n schuurpapierschijf
d Schleifpapierscheibe

* **paperhanger's brush** *GB* **→ 714**

* **paperhanger's folding rule** *GB* **→ 715**

1705 paperhanger's hammer
[dec/ham]
f marteau de tapissier
e martillo de empapelador
i martello di tappezziere
n behangershamer
d Tapezierhammer

* **paperhanger's scissors** *GB* **→ 716**

1706 parallel action pliers; parallel jaw pliers
[pli]
f pince à mâchoires parallèles
e alicate(s) de pico recto y mandíbulas paralelas
i pinza a ganasce parallele
n paralleltang
d Parallelflachzange

* **parallel cramp → 1299**

1707 parallel guide; rip fence
[pla/saf]
f guide latéral
e escuadra de guía; guía lateral
i guida laterale
n langsgeleider; parallelgeleider
d Parallelanschlag

* **parallel jaw pliers → 1706**

1708 parallel rule; parallel ruler
[mea]
f règle à parallèles; règles parallèles
e regla de paralelas
i righe parallele
n parallellineaal
d Parallellineale

1709 paring chisel
[chi]
f ciseau à décolletage
e cincel de tronzar
i scalpello per troncare
n afsteekbeitel
d Abstechbeitel

1710 paring chisel
[chi/woo]
f ciseau-fermoir long; ciseau pour

charpentiers
e cincel longo para carroceros
i scalpello lungo a taglio
n lange steekbeitel
d Treppenbauerbeitel

1711 **paring gouge; scribing gouge**
[chi]
f gouge coudée
e gubia acodada
i sgorbia ricurva
n geknikte guts
d gekröpfter Hohlbeitel

1712 **paring knife**
[for]
f boutoir; rénette; rogne-pied
e pujavante
i coltello da maniscalco
n hoefmes; veegmes
d Hufmesser; Wirkmesser

1713 **parting tool; V tool**
[chi/woo]
f gouge à rainurer en V
e gubia para ranurar en V
i sgorbia per scanalatura in V
n burein; burijn; guts met V-vormige doorsnede
d V-Hohlbeitel

* **parting tool → 1223**

1714 **paste mo(u)ld**
[fou]
f moule à pâte de charbon
e molde de pasta de carbón
i stampo a pasta di carbone
n met kool beklede vorm
d mit Kohle gefutterte Form

1715 **patent log; taffrail log**
[mea/nav]
f loch automatique; loch enregistreur à hélice
e corredera de patente; corredera helicoidal
i solcometro meccanico
n patentlog
d Patentlog

1716 **patter**
[mas]
f bouclier; guillaume de maçon
e fratás
i frettazzo
n plakspaan; spaakbord; strijkhout; strijkspaan
d Reibebrett

* **patter → 707**

1717 **patternmaker's hammer**
[car/ham]
f marteau de modeleur
e martillo de modelista
i martello di modellista
n modelmakershamer
d Modelltischlerhammer

* **patternmaker's rule → 601**

1718 **paver's hammer; roadmaker's hammer**
[civ/ham]
f marteau de paveur
e martillo de pavimentador
i martello di pavimentatore
n straatmakershamer
d Pflasterhammer

1719 **pawl; spring pawl**
[eng/mec/saf]
f cliquet d'arrêt à ressort
e trinquete con muelle
i arresto a molla
n verende pal
d Sperrklinke

1720 **pawl spring**
[eng/mec/saf]
f ressort de cliquet d'arrêt
e muelle de trinquete
i molla di arresto
n palveer
d Sperrklinkenfeder

* **peavy → 399**

1721 **pebble mill**
[min]
f broyeur à boulets
e molino de guijarros; molino de piedras de silex
i mulino a palle; mulino a sassi
n kogelmolen
d Kugelmühle

* **pebble rake → 1200**

* **pebble roller → 1201**

1722 **pedestal drilling machine**
[dri]
f perceuse à montant
e taladro de pedestal
i trapano a piedistallo
n statiefboormachine
d Ständerbohrmaschine

1723 pedometer
[mea/sur]
f piedomètre; podomètre
e cuentapasos; hodómetro; pedómetro
i pedometro
n pedometer; podometer; stappenteller
d Pedometer; Schrittzähler

* **peel → 130**

1724 peeling iron
[axe/foy]
f hache d'écorchage; tranchant coupe-écorce
e descortezador de hierro
i arnese per scortecciare
n blekbijl
d Entrindungsbeil; Loheisen; Rinden-schäler

* **peeling iron → 161**

* **peeling knife → 781**

* **peg shoulder → 11**

* **peg toothing → 1051**

* **pendulum circular saw → 822**

* **pendulum pliers → 825**

1725 pentagonal reamer
[dri/fre/rea]
f alésoir à cinq pans
e escariador de macho pentagonal
i alesatore pentagonale
n vijfhoekige ruimer
d Fünfkantaufreiber

* **perambulator → 1588**

* **percussion drill → 390**

1726 percussion drilling machine
[dri]
f perceuse à percussion
e taladro de percusión
i trapano a percussione
n klopboormachine
d Schlagbohrer

1727 percussion frequency
[dri]
f fréquence de frappe
e frecuencia de percusión
i frequenza di percussione
n slagfrequentie
d Schlagzahl

1728 percussion hammer
[ham]
f marteau de percussion
e martillo de percusión
i martello di percussione
n percussiehamer
d Perkussionshammer

1729 perforate *v* **by filing**
[far]
f percer à la lime
e perforar con la lima
i perforare colla lima
n doorvijlen
d durchfeilen

1730 permanently sharpened
[saw]
f à affûtage permanent
e de afilado permanente
i ad affilatura permanente
n blijvend scherp
d immer scharf

1731 permanent mo(u)ld
[fou]
f moule permanente
e molde fijo
i stampo permanente
n permanente vorm
d permanente Form

1732 pestle
[che]
f pilon
e majadero
i pestello
n stamper
d Keule; Pistill

1733 pick
[gen/sto]
f deux-pointes
e pico de cantería
i piccone
n piekhouweel (punt en punt)
d Zweispitz

1734 pick; pickhammer
[mas/sto]
f smille
e escoda
i piccozza
n bikhamer
d Pickhammer

* **pick → 1237**

1735 pick-axe
[gen]
f pioche à lame de hache
e alcotana; zaperpico
i piccone con lama di ascio
n houweel met bijlblad
d Breithacke

* **pick-axe → 1613**

* **pickhammer → 1734**

* **pick-up baler → 1316**

* **piercing saw → 1453**

1736 pillar drilling machine
[dri]
f perceuse à colonne
e taladro de columna
i trapano a colonna
n kolomboormachine
d Säulenbohrmaschine

1737 pillar file
[far]
f lime pilier
e lima plana paralela
i lima piatta parallela
n pilaarvijl; platstompe vijl
d flachstumpfe Werkstattfeile

1738 piloted laminate cutters; piloted laminate millers
[car/fre/wool]
f fraises à biseauter les bords avec guide
e fresas biseladoras con guía
i frese per bisellare con guida
n kantfrezen met leipen
d Kantenfräser mit Anlaufzapfen

1739 pilot ga(u)ge
[ele/meal]
f jauge-tampon à rainure circonférentielle
e calibre macho con ranura circunferencial
i calibro a tampone con canale circonferenziale
n gatkaliber met omtreksgroef
d Lehrdorn mit Umfangsrinne

1740 pilot nail
[iwa]
f clou d'attache
e clavo de fijación
i chiodo di fissazione
n hechtspijker; sloopspijker
d Haftnagel

* **pilot nail → 944**

1741 pilot pin
[gen]
f cheville de guidage
e vástago-guía
i perno di guida
n geleidepen; leipen
d Führungsstift

1742 pin; pintle
[fas/iwa]
f clou
e espiga
i spina
n pen
d Stift

1743 pin; wooden peg
[car/cpt/gen]
f cheville en bois
e clavo de madera
i chiodo di legno
n houten pen; houten spie
d Holznagel

1744 pincers; pliers; tongs
[pli]
f pince(s); tenaille(s)
e alicate(s); tenaza(s)
i pinza(e); tenaglia(e)
n tang(en)
d Zange(n)

1745 pin ga(u)ge
[meal]
f jauge-cheville
e calibrador macho; galga con extremos esféricos
i perno calibratore; perno di taratura
n meetdoorn
d Messdorn; Messzapfen

1746 pin hammer; telephone hammer
[ele/ham]
f marteau à boule; marteau de menuisier
e martillo de bola
i martello a testa tonda
n ingenieurshamer; toetshamer
d Kugelpinnenhammer

1747 pin hinge
[fas/iwa]
f charnière à clou
e espiga de bisagra
i cerniera a spina
n penscharnier
d Stiftscharnier

1748 pinion; spur wheel
[eng]

f pignon
e piñón
i pignone; rocchetto dentato
n rondsel
d Getriebe; Ritzel; Treibrad

1749 **pin reamer**
[dri/fre/rea]
f alésoir pour trous de goupille
e escariador para agujeros de pasadores
i alesatore per fori di perni
n pengatruimer
d Stiftlochreibahle; Stiftlochschälreibahle

1750 **pin spanner; sickle spanner**
[spa]
f clé à ergots
e llave de pitones
i chiave a pioli
n penhaaksleutel
d Stifthakenschlüssel

* **pintle → 1742**

1751 **pin tongs**
[wat]
f étau d'horloger
e mordaza de relojero
i mandrino per orologiai
n handbankschroefje voor horlogemakers
d Uhrmacherschraubstock

1752 **pipe clip**
[hyd/joi]
f collier de tuyau
e abrazadera para tubos; zuncho
i collarino; graffa per tubi
n buisklem
d Rohrschelle

1753 **pipe cutting apparatus; tube cutting apparatus**
[cut/hyd]
f coupe-tubes
e cortatubos
i tagliatubi
n pijpsnijder
d Rohrschneider

* **pipe wrench → 703, 1083**

1754 **pistol grip**
[cpt/dri/gen]
f poignée-revolver
e mango de pistola
i manico a pistola
n revolverhandgreep
d Pistolenhandgriff

1755 **pitchfork**
[agr]
f bident
e bidente; bieldo
i bidente; forcone
n hooigaffel
d Heugabel

1756 **pitching tool**
[chi/sto]
f ciseau large de tailleur
e cincel ancho de picapedrero
i scalpello largo di tagliapietre
n brede job
d breiter Steinhauermeissel

1757 **pitch of a plane**
[pla]
f angle fer-semelle
e ángulo hierro-base
i angolo ferro-base
n hoek beitel-zool
d Winkel Eisen-Sohle

1758 **pitch of a screw**
[mea/scr]
f pas de vis
e paso del filete
i passo di una vite
n spoed van een schroef
d Schraubengang

1759 **pitch of rivets**
[riv]
f écartement des rivets
e paso del roblonado
i spazio tra ribattini
n steek bij klinkwerk
d Nietteilung

1760 **piton; rock piton**
[iwa]
f piton
e clavija para roca
i chiodo da roccia
n stijgijzer
d Fichtelhaken; Klettereisen; Mauerhaken

1761 **piton hammer**
[ham]
f marteau d'escalade; marteau-piolet
e martillo de escalada
i martello per ascensione
n stijghamer
d Kletterhammer

* **pit saw → 537, 1543**

* **plain rubbing brick → 1205**

1762 plane *v*
[pla]
f raboter
e cepillar
i piallare
n schaven
d hobeln

1763 plane back iron; plane cap iron
[cpt/pla]
f contrefer
e contrahierro
i controferro
n keerbeitel
d Hobeldeckel

1764 plane bit; plane cutter; plane iron
[cpt/pla]
f fer de rabot
e hierro de cepillo
i ferro di pialla
n schaafbeitel
d Hobeleisen

1765 plane body; plane stock
[cpt/pla]
f fût de rabot
e caja de cepillo
i ceppo di pialla; corpo di pialla
n schaafblok
d Hobelkasten

* **plane cap iron → 1763**

* **plane cover iron → 1524**

* **plane cutter → 1764**

1766 plane handle; plane toat
[cpt/pla]
f poignée de rabot
e empuñadura de cepillo
i impugnatura di pialla
n schaafgreep; schaaftoot
d Hobelgriff

* **plane iron → 1764**

1767 plane iron sharpener
[pla]
f jauge d'affûtage du fer
e accesorio de afilado del hierro
i guida per affilatura del ferro
n beitelslijpmal
d Beitelnachschärfgerät

1768 planer attachment
[pla]
f adaptation rabot
e adaptador de cepillo
i adattamento di pialla
n schaafvoorzetstuk
d Hobelvorsatz

1769 planer file
[far/pla]
f rabot-lime
e cepillo-lima
i pialla-lima combinata
n schaafvijl; vijlschaaf
d Kombi-Hobel

1770 planer rasp
[far/pla]
f râpe à raboter
e cepillo-escofina
i pialla-raspa combinata
n raspschaaf
d Kombi-Hobel

1771 planer saw
[saw]
f scie à coupe lisse
e sierra de corte liso
i sega piallatrice
n schaafzaag
d Hobelzahnsäge

1772 planer saw blade
[cpt/saw]
f lame de scie à coupe lisse
e hoja de sierra de corte liso
i lama di sega piallatrice
n schaafzaagblad
d Hobelzahnsägeblatt

1773 planer tools
[pla/tool]
f outillage de raboteuse
e herramientas de cepillar
i utensileria di piallatrice
n schaafmachinegereedschap
d Hobelmaschinenwerkzeuge

1774 planer tools
[pla/tool]
f outillage à raboter
e herramientas de cepillar
i utensileria per piallare
n schaafgereedschap
d Hobelwerkzeuge

1775 planes
[pla]
f rabots
e cepillos
i pialle; pialletti; pialloni

n schaven
d Hobel

* **plane stock → 1765**

* **plane toat → 1766**

1776 **planing**
[pla]
f rabotage
e acepillado
i piallatura
n schaven
d Hobeln

1777 **planing bench**
[car/cpt]
f établi à raboter
e banco para cepillar
i banco per piallare
n schaafbank
d Hobelbank

1778 **planing machine**
[pla]
f raboteuse
e acepilladora
i piallatrice
n schaafmachine
d Hobelmaschine

* **planing machine headstock → 1330**

1779 **planisher; planishing hammer**
[ham/pol]
f marteau à planer
e martillo aplanador
i martello da spianare
n planeerhamer; polijsthamer
d Planierhammer; Schlichthammer

* **planting stick → 734**

* **plasterer's hawk → 1315**

* **plasterer's trowel → 1004, 1569**

* **plasterer's twitcher → 64**

1780 **plastic handle**
[cpt/gen/too]
f poignée plastique
e empuñadura plástica
i impugnatura plastica
n kunststofhandgreep
d Plastikgriff

1781 **plate ga(u)ge**
[mea]
f jauge pour tôles
e galga para chapas
i sagoma per lamiere
n blikmal
d Blechlehre

1782 **platen**
[gen]
f plateau; table de machine-outil
e mesa de máquinas-herramientas
i piano scorrevole; piastra metallica
n montagetafel
d Montagetisch

1783 **plate planishing hammer; sheet planishing hammer**
[ham/pol]
f marteau à planer la tôle
e martillo para planear la chapa
i presella per spianare la lastra
n blikhamer
d Blechschlichthammer

1784 **plate shears** *GB*; **sheet shears** *GB*; **tinman's shears** *US*
[met/she]
f cisailles à tôles; coupeur
e cizallas de chapa; tijeras para chapa
i cesoie da lamiere; cesoie da lastra; cesoie per lamierino
n blikschaar
d Blechschere

* **pliers → 1744**

* **pliers with insulated handles → 1435**

* **plier wrench → 1624**

1785 **plough; plough plane; tonguing plane**
[pla]
f bouvet à approfondir; bouvet de deux pièces; guimbarde
e cepillo para machihembrar
i incorsatoio
n veerploeg
d Federhobel; Grathobel; Karnieshobel

1786 **ploughed and tongued joint; tongue and groove joint**
[car/joi/woo]
f assemblage à rainure et languette
e empalme de ranura y lengüeta
i incastro a maschio e femmina
n messing- en groeflas
d Nut- und Federverbindung; Spundung

1787 **plough plane**
[car/pla/woo]

f bouvet femelle
e acanalador
i incorsatoio
n ploegschaaf; veerploeg
d Nothobel

* **plough plane → 1785**

1788 plug
[gen]
f bouchon; tapon
e tapón; taco
i tappo
n plug
d Dübel

1789 plug box
[fas/iwa]
f étui à bouchon; étui à tapon
e casa de tapones
i cassetta a tappe
n pluggendoos
d Dübelschachtel

1790 plug ga(u)ge
[ele/mea]
f calibre à tampon cylindrique
e calibre de agujeros
i calibro a tampone
n gatkaliber; penkaliber
d Lehrdorn

* **plug ga(u)ge → 1441**

1791 plugging chisel; seam drill
[chi/sto]
f tamponnoir
e escarbador
i scalpello per inserire tappi
n plugbeitel
d Dübelmeissel; Dübellochmeissel

* **plug tap → 297**

1792 plumb bob
[car/mas/sur]
f fil à plomb
e cuerda de plomada
i filo a piombo
n schietlood
d Senklot

1793 plumber's hammer
[ham/hyd]
f marteau de plombier
e martillo de plomero
i martello d'idraulico
n loodgietershamer
d Klempnerhammer

1794 plumb level; plumb rule
[mas/mea/sur]
f niveau de maçon; niveau vertical
e nivel de albañil; nivel de plomada
i livella orizzontale e verticale
n metselaarswaterpas
d Horizontal-Vertikal-Wasserwaage; Setzwaage

* **plummet level → 1579**

1795 plunge grinding; profile grinding
[gri]
f rectification de profils
e rectificado con plantilla
i rettifica con mola sagomata
n profielslijpen
d Profilschleifen

* **plunge grinding → 1090**

1796 plunger
[civ/eng]
f piston
e émbolo
i pistone; stantuffo
n zuiger
d Stempel

* **pneumatic drill → 390**

1797 pneumatic drilling machine
[dri]
f perceuse à air comprimé
e taladro neumático
i trapano pneumatico
n pneumatische boor
d Pressluftbohrmaschine

1798 pneumatic ga(u)ge
[mea]
f comparateur à air comprimé
e comparador de aire comprimido
i comparatore ad aria compressa
n luchtdrukvergelijker
d Luftdruckvergleicher

1799 pneumatic hammer
[eng/ham]
f marteau à air comprimé; marteau pneumatique
e martillo neumático
i martello pneumatico
n luchtdrukhamer
d Presslufthammer

* **pneumatic hammer → 38**

1800 pneumatic pick
[civ/engl]
f pic pneumatique
e picador neumático
i piccone pneumatico
n luchtdrukhouweel
d Pressluftpickel

* **pneumatic power hammer → 38**

1801 pneumatic reamer
[dri/fre/real]
f alésoir pneumatique
e escariador neumático
i alesatore pneumatico
n luchtdrukruimer
d Pressluftreibahle

1802 pneumatic riveter
[rivl]
f riveuse à air comprimé
e remachadora de aire comprimido
i chiodatrice pneumatica
n luchtdrukklinker
d Druckluftnietmaschine

1803 pneumatic riveting hammer
[ham/rivl]
f marteau pneumatique rivoir
e martillo neumático de remachar
i martello pneumatico per chiodare; martello pneumatico per ribadire
n pneumatische klinkhamer
d Druckluftniethammer

1804 pneumatic screwdriver; power screwdriver
[scdl]
f tournevis pneumatique; visseuse
e destornillador neumático
i cacciavite pneumatico
n pneumatische schroevedraaier
d Pressluftschrauber

1805 pneumatic tools
[genl]
f outils pneumatiques
e herramientas neumáticas
i arnesi pneumatici; utensileria pneumatica
n luchtdrukgereedschap
d Druckluftwerkzeuge

1806 podauger
[dril]
f tarière à rainure longitudinale
e barrena de canaleta recta
i macchina a tortiglione e scanalatura retta; trivella a scanalatura retta
n avegaar met overlangse groef
d Stangenschlangenbohrer mit geradliniger Nute

1807 point
[chi/stol]
f ciseau pointu
e cincel de punta
i scalpello a punta
n puntbeitel
d Spitzmeissel

* **pointed chisel → 583**

1808 pointed countersink
[dri/real]
f alésoir-fraise à pointe
e avellanador de punta
i accecatoio a punta
n puntverzinkfrees
d Spitzensenker

1809 pointed file
[farl]
f lime pointue
e lima terminada en punta
i lima a punta
n spitse vijl; spitsvijl
d Spitzfeile

1810 pointed hammer; stippling hammer
[ham/stol]
f marteau à pointe
e martillo de picadrero; martillo para picar
i picchierello
n bosseerhamer; punthamer; spitshamer
d Bossierhammer; Spitzhammer

* **pointed trowel → 1153**

1811 pointer; raker
[masl]
f grattoir; racloir
e rascador para rejuntar
i grattino
n voegkrabber
d Kratzeisen

* **pointer → 1467**

1812 pointing
[cle/masl]
f grattage
e rascadura
i raschiatura
n voegen uitkrabben
d Fugenkrätzen

1813 pointing iron
[mas]
f fiche de maçon
e paletín de rejuntar
i nettatoio
n voegijzer
d Fugeneisen; Stukkateureisen

1814 pointing trowel
[mas]
f truelle lisseuse
e paleta a palustre
i cazzuola a lama rombica
n pleistertroffel
d Putzerkelle; Reparaturkelle

1815 point of a file
[far]
f pointe de lime
e punta de lima
i punta di lima
n punt van een vijl
d Feilspitze

1816 poker
[hou]
f pique-feu
e atizador
i attizzatoio
n pook
d Feuerhaken; Schürhaken

1817 polishers
[pol]
f lustreuses; machines à polir
e pulidoras
i lucidatrici
n polijstmachines
d Poliermaschinen

1818 polisher-sander; sander-polisher
[pol/san]
f lustreuse-ponceuse
e pulidora-lijadora
i smeriglio-lucidatrice; smeriglio-pulitrice
n schuur- en polijstmachine
d Schleif- und Poliermaschine

1819 polishing
[pol]
f brunissage; polissage
e bruñido; pulido; pulimento
i brunitura; lucidatura; pulitura
n polijsten
d Blankschleifen; Polieren

1820 polishing block
[pol]
f cale à poncer
e bloque para pulido
i blocco pulitore
n schuurblok
d Polierklotz; Schleifklotz

1821 polishing bonnet
[cpt/san]
f bonnette de polissage
e funda pulidora
i cuffia pulitrice
n polijsthoes
d Polierhaube

1822 polishing file
[far/pol]
f polissoir
e lima pulidora
i lima lucidatrice
n schuurstaal
d Polierfeile; Polierstahl

1823 polishing head; polishing lathe
[lat/pol]
f poupée de tour à polissage
e cabezal pulidor; torno para pulidores
i testa motrice; lucidatrice
n polijstkop
d Polierkopf

1824 polishing steel
[lea/pol]
f polissoir de cuir
e pata de cabra
i liscia
n polijststaal
d Glätteisen; Glättstein

1825 polishing steel
[pol]
f polissoir en acier
e bruñidor; pulidor de acero
i raschietto per pulire
n bruineerijzer; bruineerstaal; poleerstaal; polijststaal
d Polierstahl

1826 polishing stick
[pol]
f tas à planer
e esmerilador de madera
i lisciatoio
n polijststok
d Polierstift

1827 polishing stick
[lea/pol]
f astic
e costa
i bisegolo; bussetto

n gladhout; likbeen; likhout
d Glättholz

1828 **poll-adze**
[axe/car/coo/wool]
f herminette à tête plate
e azuela de cabeza plana
i ascia a testa piatta
n dissel met platte kop
d Flachkopfdexel

1829 **pontie; pontil; puntee; punty**
[gla]
f pontil
e pontil; puntel
i puntello
n hechtijzer; punteerijzer
d Hefteisen

* **pope's head → 444**

* **pop riveter → 257**

* **porcupine → 1178**

1830 **portable circular saw**
[saw]
f scie circulaire portatile
e sierra circular portátil
i sega circolare portatile
n draagbare cirkelzaag
d tragbare Kreissäge

* **portable lamp → 1280**

1831 **positioner**
[cpt/dri]
f lunette
e posicionador; situador en posición
i apparecchio per messa in posizione
n geleidingsring
d Vorsatzhalter

1832 **potato dibble**
[agr/nut]
f plantoir de pommes de terre
e sembradora de patatas
i foraterra per patate
n aardappelboor; aardappelkruk; aardappelpoothout
d Kartoffelsetzholz

1833 **potato digging fork**
[agr/nut]
f arracheur de pommes de terre; fourche à arracher les pommes de terre
e arrancadora de patatas; horquilla para arrancar patatas
i forca scavapatate
n aardappelrooier; aardappelrooivork
d Kartoffelerntegabel; Kartoffelgrabegabel; Kartoffelschaufel

* **potato hoe → 398**

1834 **potato loading fork**
[agr/nut]
f fourche à pommes de terre
e horquilla para cargar patatas
i forca da patate
n aardappelvork
d Kartoffelkralle

1835 **power hammer**
[ham]
f marteau autocompresseur
e martillo neumático de estampar
i maglio pneumatico per la forgiatura su stampo
n luchtdrukhamer voor perswerk
d Luftgesenkhammer

1836 **power hammer**
[ham]
f marteau-pilon
e martillo pilón; martinete
i maglio meccanico
n smeedhamer
d Schmiedehammer

1837 **power jig saw; sabre saw**
[saw]
f machine alternative à découper
e sierra eléctrica de recortar
i sega elettrica a volgere
n elektrische decoupeerzaag
d elektrische Dekupiersäge

1838 **power saw**
[saw]
f scie à moteur
e sierra mecánica; sierra motorizada
i sega meccanica
n motorzaag
d Motorsäge

* **power screwdriver → 1804**

1839 **precision aids**
[pri]
f outillage auxiliaire de précision
e equipo auxiliar de precisión
i apparecchiatura ausiliaria di precisione
n precisiehulpmiddelen
d Präzisionshilfsmittel

1840 **precision drilling machine**
[dri]

f perceuse à précision; perceuse de calibres
e taladro a calibre; taladro de precisión
i trapano di precisione
n fijnboormachine
d Lehrenbohrmaschine

1841 precision file
[far]
f lime de précision
e lima de precisión
i lima di precisione
n precisievijl
d Mechanikerfeile; Präzisionsfeile

1842 precision tools
[gen]
f outils de précision
e herramientas de precisión
i utensileria di precisione
n precisiegereedschap
d Präzisionswerkzeuge

1843 precision water level
[mea/sur]
f niveau à bulle d'air de précision
e nivel de burbuja de precisión
i livella a bolla di aria di precisione
n aswaterpas
d Wasserwaage zum Ausrichten von Wellen

1844 press tool
[tool]
f matrice pour presses
e estampa; traquel de prensa
i matrice di pressa
n persmatrijs
d Pressmatrize

1845 pressure roll
[cpt/gri]
f rouleau-presseur
e cilindro prensador
i rullo di pressione
n drukrol
d Druckrolle

1846 profile chisel
[chi/met/woo]
f ciseau à profiler
e cincel de perfilar
i scalpello per profilare; scalpello per sagomare
n profielbeitel
d Profilbeitel

1847 profile cutting shears
[met/sci]
f cisaille à profiler
e cizalla para perfiles
i cesoia per profilare
n figuurblikschaar; figuurschaar; handblikschaar; knobbelschaar; gatblikschaar
d Figuren-Blechschere

1848 profiled joint
[car/joi/woo]
f assemblage profilé
e asembladura perfilado
i encastro profilato
n geprofileerde verbinding
d profilierte Verbindung

*** profile grinding → 1090, 1795**

1849 profile iron
[gen/iwa]
f fer profilé
e hierro perfilado
i ferro profilato; ferro sagomato
n façonijzer
d Fassoneisen

1850 profile wire
[gen/iwa]
f fil profilé
e filo perfilado
i filo profilato; filo sagomato
n façondraad
d Fassondraht

1851 profiling
[gen]
f profilage
e perfilado
i profilatura
n profileren
d Profilierung

1852 progressive press tool
[tool]
f matrice progressive pour presses
e estampa progresiva
i matrice di pressa progressiva
n meelopende persmatrijs
d mitlaufende Pressmatrize

1853 progressive toothing
[saw]
f denture à pas croissant
e dentadura de paso de crecimiento
i dentatura a passo ad accrecimento
n vertanding met toenemende steek
d progressive Zahnung

1854 protractor
[mea]
f rapporteur; transporteur

e transportador
i rapportatore
n graadboog; graadmeter; hoekmeter; transporteur
d Transporteur; Winkelmesser

1855 **pruning knife**
[agr/foy/hor]
f serpe; serpette
e podadera
i coltello da rimonda
n heep; hiep; snoeimes
d Asthippe; Astkniepe

1856 **pruning saw**
[foy/hor/saw]
f scie d'élagage
e sierra de leñador
i potatoio; roncola
n snoeizaag
d Zweigsäge

* **pruning saw blade → 1133**

1857 **pruning shears; secateur**
[agr/foy/hor]
f sécateur
e tijeras de podar
i forbici da potare; forbici da rimonda
n snoeischaar
d Rosenschere

1858 **puller for inside operation**
[gen/pull]
f extracteur à prise intérieure
e extractor para trabajos interiores
i estrattore per lavori interni
n binnentrekker; trekker voor inwendig trekwerk
d Innenabzieher

1859 **puller for outside operation**
[gen/pull]
f extracteur à prise extérieure
e extractor para trabajos exteriores
i estrattore per lavori esterni
n buitentrekker; trekker voor uitwendig trekwerk
d Aussenabzieher

1860 **pullers**
[pull]
f extracteurs
e extractores
i estrattori
n trekkers
d Abzieher

1861 **pulley**
[cpt/eng]
f poulie
e polea
i puleggia
n katrol
d Rolle

* **pulley → 1162**

1862 **pulley block**
[cpt/tra]
f bloc de palan; moufle
e motón
i bozzello
n katrolblok
d Block (Flaschenzug)

* **pulley block → 259**

1863 **pulling claws**
[cpt/pull]
f griffes d'extraction
e garras de extracción
i griffe d'estrazione
n trekarmen
d Abzughaken

1864 **pulling yoke**
[pull]
f joug d'extraction
e yunta de extracción
i giogo d'estrazione
n trekjuk
d Abziehfutter

1865 **punch**
[gen]
f poinçon
e punzón
i punzone
n pons
d Lochstempel; Punze; Stanzmesser

1866 **punching hammer**
[ham/per]
f marteau à poinçonner
e martillo punzonador
i martello punzonatore
n ponshamer
d Lochhammer

1867 **punch movement**
[cut]
f mouvement du poinçon
e movimiento del útil de estampado
i movimento del punzone
n ponsbeweging
d Stanzmesserbewegung

1868 punch pliers
[pli]
f pince emporte-pièce revolver
e sacabocado(s)
i pinza a fustella
n holtang; perforeertang
d Lochzange

1869 punner; punning tool
[civ/cpt]
f dame; demoiselle; hie; pilon
e herramienta de apisonamiento
i mazzaranga
n stamper
d Handramme; Rammblock

* **puntee → 1829**

* **punty → 1829**

1870 push-pull steel tape
[meal]
f mètre à ruban en acier
e metro de cinta de acero
i metro di nastro di acciaio
n stalen rolbandmaat
d Stahlbandmass

1871 push-pull tape
[meal]
f mesure à ruban à rouleau; mètre à ruban à rouleau
e cinta métrica de rodillo
i nastro metrico in rotolo
n rolbandmaat
d Rollbandmass

1872 putty knife
[dec]
f couteau à mastiquer
e espátula para la masilla
i coltello da mastice
n stopmes
d Kittmesser

Q

1873 quadrangular file; square file
[far]
f lime carrée
e lima cuadrada; lima rectangular
i lima quadrangolare
n vierkante vijl
d Vierkantfeile

1874 quadrant dividers
[mea]
f compas à pointes sèches; compas droit à pointes
e compás de divider; compás de puntas secas
i compasso a cuadrante
n steekpasser
d Stechzirkel; Teilzirkel

1875 quarter sawing; rift sawing
[foy/saw]
f sciage de long en quatre parts
e aserrado por cuartos
i taglio di un tronco in quattro parti nel senso della grana
n zagen van boomstammen in vier delen in de lengterichting
d in vier gleiche Teile aufsägen

1876 quick-acting vice; quick-grip type of vice
[cpt/vic]
f étau à action rapide
e tornillo de banco de apriete rápido
i morsa ad azione rapida
n snelklemmende bankschroef
d schnellklemmender Schraubstock

1877 quick-change drill chuck
[cpt/dri]
f mandrin à change rapide
e portabroca de cambio rápido
i mandrino a scambio rapido
n sneluitwisselbare boorhouder
d Schnellwechselbohrfutter

1878 quick-change lever; quick-release toggle
[cpt/saw]
f tendeur à remplace rapide
e palanca de apriete rápido
i leva a forcello per scambio rapido
n zaagsnelspanner
d Schnellspanner

1879 quick change self-gripping chuck
[cpt/dri]
f mandrin à autoserrage et changement rapide
e mandril autosujetador de cambio rápido
i mandrino automatico a scambio rapido
n sneluitwisselbare zelfklemmende boorhouder
d Schnellwechselspannfutter

1880 quick gouge
[car/chi]
f gouge à cambrure forte
e gubia de combeo grande
i sgorbia ad incurvatura grande
n guts met grote welving
d Hohlbeitel mit grosser Wölbung

* **quick-grip type of vice → 1876**

1881 quick-motion screw
[iwa/scr]
f vis à grand pas; vis à pas rapide
e tornillo de paso grande
i vite a passo rapido
n schroef met grote spoed
d Schraube mit grosser Steigung

* **quick-release toggle → 1878**

1882 quick-spiral twist drill
[dri]
f mèche hélicoïdale a grand pas
e macho de espiral de paso grande
i punta elicoidale a gran passo
n spiraalboor met grote spoed
d Spiralbohrer mit grosser Steigung

1883 quirk float
[mas]
f truelle à profiler
e llana para perfilar
i cazzuola per modanature
n profielspaan
d Profilkelle

R

1884 **rabbeted lock**
[hou/saf]
f serrure à mortaiser
e serradura embutida; serradura encajada
i serratura incassata
n inbouwcilinderslot
d Einbauschloss

1885 **rabble**
[met/min]
f râble
e barra batidora
i sbarra per rimescolare
n roerstaaf
d Rührstange

* **race puller → 141**

1886 **rack saw**
[car/saw]
f scie à denture grosse
e sierra de corte ancho; sierra de dientes anchos
i sega a denti grossi
n groftandige zaag
d Grobsäge; grobverzahnte Säge

1887 **radial-arm drilling machine**
[dri]
f perceuse radiale universelle
e taladro radial universal
i trapano radiale universale
n universele radiaalmachine
d Radialbohrmaschine mit stufenlosem, servomotorgesteuertem Drehzahlgetriebe

1888 **radial arm saw**
[saw]
f scie mécanique à raccourcir
e sierra mecánica para acortar
i sega meccanica per scorciare
n afkortzaagmachine; radiaalzaag
d Abkürzsägemaschine

1889 **radial drilling machine**
[dri]
f perceuse radiale
e taladro radial
i trapano radiale
n radiale boormachine
d Radialbohrmaschine

* **radial square → 449**

* **radiator brush → 56, 646**

* **radio pliers → 1544**

1890 **radius ga(u)ge**
[mea]
f appareil de mesure de rayons
e medidor de radios
i misuratore di raggi
n straalmeter
d Radienlehre

* **rafter square → 417**

1891 **rag bolt**
[iwa]
f boulon barbelé
e perno arponado; perno de anclaje; perno para empotrar
i bullone a mazzetta
n bout met weerhaken
d Bartbolzen

* **rag bolt → 766, 1094, 1448**

1892 **rail bender**
[ben/tra]
f cintreuse de rails
e prensa curvacarriles
i apparecchio per curvare le rotaie
n railbuiger; railbuigmachine
d Schienenbiegemaschine

1893 **rail saw**
[saw/tra]
f scie à rails
e sierra de carriles
i sega per rotaie
n railzaag
d Schienensäge

1894 **rake**
[agr/hor]
f râteau
e rastrillo
i rastrello
n hark
d Harke; Rechen

1895 **rake**
[hou/tra]
f raclette
e hurgón
i rastrello da stufa
n slakkesteker; vuurhaak
d Schüreisen

1896 **rake**
[mas]
f râble de maçon
e barra batidora de albañil

i mestatoio; sbarra per rimescolatura di muratore
n metselaarsroerstang
d Maurerrührstange

* **raker → 1811**

1897 raker set
[cpt/saw]
f denture à interchangement alterné
e disposición de dientes troscados alternativamente con un tercero recto
i dentatura a intercambio alternato
n tandverzetting
d verstellte Zahnung

1898 ram; tup
[ham]
f masse tombante; mouton
e mazo batente
i mazza battente; mazza cadente
n beer van stoomhamer
d Hammerbär

1899 rammer
[fou]
f batte de mouleur; fouloir
e bate de moldeador
i mazza
n stamper
d Stampfer

1900 ramshorn handle of a plane
[cpt/pla]
f corne de rabot
e cuerno de cepillo
i corno di pialla
n hoorntje van een schaaf
d Hobelgriff

1901 ranging rod
[sur]
f perche de mire
e barra de mira
i asta di mira; palina
n nivelleerstok; richtbaak
d Fluchtstab; Visierstab

1902 rapid glue press; spring cramp
[car/cpt]
f presse rapide à coller
e prensa rápida de encolar
i pressa rapida da incollatura
n snellijmtang
d Schnelleimzwinge

1903 rasp
[far]
f râpe
e escofina
i raspa
n rasp
d Raspel

1904 rasp cut
[far]
f piqûre de râpe
e picadura de escofina
i taglio di raspa
n puntkap; raspvertanding
d Raspelhieb

1905 ratchet drilling apparatus
[dri]
f vilebrequin à rochet
e carraca de taladrar; chicharra; taladro de carraca
i trapano a cricco
n ratelboor
d Bohrknarre

1906 ratchet mechanism
[eng/gen/scd]
f cliquet; rochet
e carraca; mecanismo de trinquete
i cricco
n ratelmechanisme
d Knarre

1907 ratchet screwdriver
[scd]
f tournevis à cliquet
e destornillador con carraca
i cacciavite a cricchetto
n ratelschroevedraaier
d Knarrschraubendreher

1908 rat-tail file
[far]
f queue de rat
e cola de rata
i lima a coda di topo
n kurkvijl; rattestaart
d Rattenschwanz

1909 raw-hide hammer; soft faced hammer; soft faced mallet
[ham]
f marteau à face en cuir vert
e mazo de cabeza con cuero verde
i martello con testa di pelle gregge
n kurashamer; ruwhuidhamer
d Rohhauthammer

1910 reamer
[dri/fre/real]
f alésoir
e avellanador; escariador

i accecatoio
n ruimdoorn; ruimer
d Reibahle

1911 reaping hook; sickle
[agr/hor]
f faucille
e falce; guadaña; hoz
i falce
n sikkel
d Sichel

1912 rebate bit; rebate cutter
[car/fre/woo]
f fraise à feuillure
e fresa de machihembrar
i fresa per scanalare
n sponningsfrees
d Falzfräser

1913 rebate joint
[car/joi/woo]
f assemblage à encastrement; assemblage à feuillure
e ensambladura a rebajo; asembladura quebrada
i incastro incassato
n rabatnaadverbinding
d Einfalzung; Einfügung

1914 rebate plane; shoulder plane
[pla]
f guillaume; jablière; jabloir
e cepillo rebajador; guillame
i pialletto; sponderuola
n boorschaaf; rabatschaaf; rijtuigschaaf; sponningschaaf
d Absatzhobel; Falzhobel

* **reciprocating saw → 175**

* **reciprocating saws** *GB* **→ 1125**

1915 reducer; reduction socket
[cpt/hyd]
f manchon de réduction
e reductor
i riduttore; riduzione
n verloopstuk
d Reduktionsmuffe

1916 reducing pipe-joint; reducing socket
[cpt/dri/hyd]
f douille de réduction
e casquillo cónico; casquillo de reducción
i bussola di riduzione
n reduceerkop
d Reduzierhülse

1917 reduction cal(l)ipers
[mea]
f compas de réduction
e compás de reducción
i compasso di riduzione
n reductiepasser
d Reduktionszirkel; Teilzirkel

* **reduction socket → 1915**

1918 reed roof hook; straw roof hook; thatched roof hook
[cov/hou]
f crochet pour toit de chaume; crochet pour toit de roseau
e garfio para techado de caña; garfio para techado de junja; garfio para techado de paja
i gancio per il tetto di canna; gancio per il tetto di paglia
n dekhaak
d Dachdeckenhaken; Rohrdachhaken; Strohdachhaken

* **reference gage** *US* **→ 1581**

1919 refuse shovel
[cle]
f pelle à ordures
e cogedor para basuras
i palla per rifiuti
n puinschop
d Kehrichtschaufel

1920 registered mortise chisel
[chi/sto]
f bédane à poids élevé
e bedano pesado
i bedano pesante
n zware schietbeitel
d schwerer Lochbeitel

1921 rehardened hook type of toothing
[saw]
f denture à crochets retrempée
e dentadura de ganchos retemplada
i dentatura a ganci ritemperata
n nageharde haakvormige vertanding
d nachgehärtete KV-Verzahnung

1922 reinforced backsaw
[saw]
f scie à dos renforcé
e sierra de dorso reforzado
i sega a dorso rinforzato
n kapzaag met versterkte rug
d Rückensäge mit verstärktem Rücken

1923 repointing
[mas]
f réparation de jointures
e reparación de junturas
i riparazione di giunti
n repareren van voegen
d Reparieren von Fugen

1924 resaw
[foy/saw]
f débitage
e subdivisión de troncos
i suddivisione di tronchi
n verzagen
d Zersägen

1925 resharpenable
[saw/sha]
f réaffûtable
e reafilable
i riaffilabile
n naslijpbaar
d nachschärfbar

1926 resharpenable tungsten carbide-tipped toothing
[saw/sha]
f denture à plaquettes en carbure de tungstène réaffûtable
e dentadura con cuchillas de carburo de tungsteno reafilable
i dentatura al carburo di tungsteno riaffilabile
n naslijpbare hardmetaal-vertanding
d nachschärfbare hartmetall-bestückte Verzahnung

1927 resinoid-bonded abrasive
[gri/san]
f abrasif à agglomérant résinoïde
e abrasivo ligado con resina sintética
i abrasivo ad impasto di resina sintetica; abrasivo collato con resina sintetica
n met kunsthars gebonden slijpmiddel
d kunstharzgebundenes Schleifmittel

1928 retort
[che]
f cornue; retort
e retorta
i storta
n retort
d Retorte

1929 retractable guard
[saw]
f carter escamotable
e guarda retractable
i riparo retrattile
n uitklapbare bescherming
d abklappbare Schutzhaube

1930 reversible handle
[cpt/saw]
f manche réversible
e mango reversible
i manico reversibile
n omkeerbaar handvat
d umlegbarer Griff

1931 reversible pulling claws
[cpt/pul]
f griffes réversibles
e garras reversibles
i griffe rovesciabili
n omkeerbare trekhaken
d umkehrbare Abzughaken

* **revolving die hammer → 1570**

1932 revolving head punch
[per]
f pince emporte-pièce revolver
e punzonadora revólver; sacabocado(s) revólver
i pinza a fustella tipo a revolver
n revolverponstang
d Revolverlochzange

* **ribbed sole → 627**

1933 ribbon-flame burner
[bur]
f lance à brûler platte
e quemador de llama plana
i bruciatore a fiamma piatta
n brander met platte vlam
d Flachbrenner

1934 ribbon saw
[saw]
f scie à ruban
e sierra de cinta
i sega a nastro
n lintzaag
d Bandsäge

1935 riddle
[cle/min]
f claie; crible
e criba de sacudidas
i crivello
n grove draadzeef
d grobes Drahtsieb

1936 ridging
[agr/hor]
f buttage

e aporcamiento
i rincalzatura
n aanaarden
d Anhäufung

1937 ridging hoe
[agr/hor]
f butteuse; rayonneur butteur
e aporcador
i zappa rincalzatrice
n aanaardhak
d Häufelhacke

1938 riffler
[far]
f rifloir
e lima encurvada
i raspino per curve
n riffelvijl
d Riffelfeile

* **rift sawing → 1875**

1939 rim lock; straight lock
[iwa/saf]
f serrure à bosse
e cerradura recercada
i serratura a cassetta
n oplegslot
d Kastenschloss

* **rim plane → 480**

1940 rim tempered
[ham/met]
f à rebord trempé; durci par revenu
e endurecido por revenida
i indurito per rinvenimento
n door ontlating gehard; gehard en ontlaten
d anlassgehärtet

1941 ring bolt
[fas/iwa]
f boulon à œillet
e macho de anillo
i caviglia ad anello
n ringbout
d Ringbolzen

1942 ring chuck
[cpt/dri/gen]
f mandrin annulaire
e mandril de anillo
i mandrino ad anello
n ringvormige houder
d Ringfutter

1943 ring ga(u)ge; ring size ga(u)ge
[gas/mea]
f jauge conique
e calibre para anillos; lastra de medidas
i fuso per misurare anelli
n dolk; ringstok
d Ringlehre; Ringmesser; Ringstock; Rinst

* **ring ga(u)ge → 960**

1944 ring rounding tool; triblet
[gas]
f triboulet
e lastra de ajuste
i spira per arrotondare anelli
n triboulet
d Richtkegel; Ringriegel; Rundholz

* **ring size ga(u)ge → 1943**

1945 ring spanner
[spa]
f clé polygonale
e llave de estrella
i chiave a stella
n ringsleutel
d Ringschlüssel

1946 ring-type collar grip
[cpt/dri]
f poignée-collier annulaire
e empuñadura anular de collar
i impugnatura anulare a collare
n ringvormige klemzijhandgreep
d ringförmiger Schellengriff

1947 rip *v*
[saw]
f scier de long; scier en long
e aserrar en la dirección de la fibra
i segare per il lungo
n met de draad mee zagen; schulpzagen
d in der Richtung der Längsfasern sägen

1948 rip blade
[cpt/saw]
f lame de scie de long
e hoja de sierra al hilo
i lama di sega per il lungo
n zaagblad met wolfstanden
d Langsägeblatt

* **rip fence → 1497, 1707**

1949 ripper
[cov/iwa]
f tire-clous de couvreur
e sacaclavos de tejador
i cacciachiodo di copritetti
n leidekkersspijkertrekker
d Dachdeckernageleisen

* **ripper saw → 537**

1950 **ripping chisel**
[chi]
f ciseau à démolition; pince de déballeur; pince monseigneur
e cincel de demolición; pie de cabra
i piede di porco; scalpello a demolire
n sloopbeitel
d Abbruchstange; Geissfuss

1951 **rip saw; whip saw**
[saw]
f scie à refendre; scie de long
e sierra al hilo
i sega per il lungo
n kraanzaag; schulpzaag
d Langsäge

* **rip saw → 537**

1952 **rising and falling saw**
[saw]
f scie circulaire à bas et en haut
e sierra de movimiento de subida y bajada
i sega regolabile
n instelbare cirkelzaag
d einstellbare Kreissäge

* **river → 535**

1953 **rivet**
[iwa/riv]
f rivet
e remache
i chiodo; ribattino; rivetta
n klinknagel
d Niet; Niete

1954 **rivet buster; rivet chisel**
[chi/met]
f tranche matoir
e uñeteador de remaches
i scalpello per eliminare ribattini
n klinknagelbeitel
d Nietkopfquetscher

1955 **riveted joint**
[joi/riv]
f jonction par rivets
e junta remachada
i giunto inchiodato
n geklonken verbinding
d Nietverbindung

* **riveter → 257**

1956 **rivet extractor**
[riv]
f chasse-rivets
e embutidor de remaches
i cacciaribattini; estrattore di ribattini
n klinknageltrekker
d Nietenzieher

1957 **rivet header; rivet head forming tool**
[riv]
f outil pour former les têtes de rivets
e encabezadora de remaches
i arnese per formare le teste di ribattini
n holputs; klinknagelkopvormer
d Nietenkopfmacher

1958 **rivet hole reamer**
[dri/fre/rea/riv]
f alésoir pour trous de rivet
e escariador de agujeros de remache
i alesatore per fori di ribattino
n klinknagelgatruimer
d Nietlochreibahle

1959 **riveting hammer**
[ham/riv]
f marteau à riveter; marteau-riveur; rivoir
e martillo de remachar
i martello per ribadire
n klinkhamer
d Niethammer

* **riving axe → 1110**

* **roadmaker's hammer → 1718**

1960 **rock drill**
[dri/min/sto]
f perforatrice de rocher
e perforadora para roca
i perforatrice per roccia
n rotsboor; rotsbreker
d Bohrknarre; Gesteinsbohrer

1961 **rock drilling bit** *GB*; **roller drilling bit** *US*
[dri/min/sto]
f trépan à molettes
e barrena de ballesta
i perforatore a ruota
n kneusbeitel; rolbeitel
d Rollenbohrer

* **rock piton → 1760**

1962 **roller bearing**
[eng/iwa]
f appui à rouleaux; palier à rouleaux
e rodamiento de rodillos
i cuscinetto a rulli

n rollager
d Rollenlager; Walzenlager

* **roller bit → 797**

* **roller drilling bit** *US* **→ 1961**

1963 **roman ogee bit; roman ogee cutter**
[car/fre/woo]
f fraise pour doucine renversée
e fresa para ojivas romanas
i fresa per ogive romane
n verkeerd-ojieffrees
d Fräser für romanische Spitzbogen

1964 **roof hook**
[cov/hou]
f crochet de couvreur
e gancho del techado
i gancio per la scaletta; uncino per la scaletta
n dakhaak
d Dachhaken

* **roofing square → 417**

1965 **roof ladder**
[cov/dec]
f échelle de couvreur; échelle à toiture
e escalera de techado
i scala da tetto; scala sul tetto
n dakladder
d Dachleiter

1966 **rope pulley**
[eng/tra]
f poulie à corde
e motón de cable; polea de cable
i puleggia per fune
n snaarschijf
d Seilkloben; Seilscheibe

1967 **rose countersink**
[dri/rea]
f alésoir-fraise champignon
e avellanador de roseta
i accecatoio a rosetta
n verzinkfrees met meerdere geslepen freestanden
d Krauskopf

* **rotary burs → 372**

1968 **rotary drier**
[met/min]
f séchoir rotatif; tambour de séchage
e secador rotativo
i essiccatoio rotante
n droogcentrifuge; droogtrommel
d Trockenschleuder

1969 **rotary drills**
[dri]
f perceuses; perforatrices
e perforadoras; taladradoras
i perforatrici; trapani
n boormachines
d Bohrmaschinen

* **rotary milled files and rasps → 372**

1970 **rotary wire brushes; wire rotary brushes**
[cle]
f brosses métalliques de rotation
e cepillos de alambre giratorios
i spazzole metalliche rotatorie
n roterende staalborstels
d rotierende Drahtbürsten; Schaftrundbürsten

1971 **rotating garden sprinkler**
[hor]
f arroseur rotatif
e aspersor giratorio
i irrigatore giratorio
n cirkelsproeier
d Kreissprenger

* **rough file → 84**

1972 **rough grinding**
[gri/san]
f ébarbage
e desbarbado; rebarbado
i sbavatura; sgrossatura
n afbramen
d Abgraten von Stahl

1973 **rough grinding wheel**
[cpt/gri]
f disque à ébaucher; disque affûtage ébauche
e muela rectificadora gruesa
i mola per sgrossare
n grofslijpschijf
d Grobschleifscheibe

1974 **roughing tool**
[met/pla/pol]
f outil pour dégrossir
e herramienta de desbastar
i utensile per sgrossare
n afbraamgereedschap
d Schruppwerkzeug

1975 **round**
[car/pla]

f doucine
e cepillo redondeador
i pialla per cornici
n bandschaaf
d Rundhobel

1976 round; rung; spoke
[fir/gen/saf]
f barreau; échelon
e peldaño
i piolo
n sport
d Sprosse

1977 round awl; straight cylindrical awl
[dri/lea]
f alène ronde
e lezna redonda
i punteruolo a lama tonda
n priem met rond lemmet
d Riemenahle; Rundahle

1978 round beck iron
[cpt/for]
f tas à tête ronde; tasseau à tête ronde; tasseau rond
e tas redondo
i tassello tondo
n ronde tas
d runder Tasso

1979 round belt pliers
[pli]
f pince à perforer la courroie
e sacabocado(s) para correa
i pinza a fustella per cinghia
n riemperforeertang
d Riemenlochzange

1980 round brush
[dec]
f brosse ronde
e brocha redonda
i pennello rotondo
n ronde kwast
d Ringpinsel; Rundpinsel

1981 round-edge slipstone; tapered slipstone
[chi/gri/sha]
f pierre à affûter gouges
e piedra de amolar para gubias
i pietra da affilare per sgorbie
n gutsensteen; tapse wetsteen
d Hohlmeisselstein

1982 round file
[far]
f lime ronde
e lima redonda
i lima rotonda; lima tonda
n ronde vijl
d Rundfeile

1983 round flat stake
[for]
f rognon rond-plat
e tas redondo-plano
i tassello rotondo-piatto
n staak met ronde platte baan
d Faust (Bahn flachrund)

1984 round-head rivet; snap-head rivet
[iwa/riv]
f rivet à tête ronde
e remache de cabeza redonda
i ribattino a testa tonda
n klinknagel met ronde kop
d Rundkopfniete

* **rounding adze → 1381**

1985 rounding cutter
[fre]
f fraise pour arrondir le arêtes de dents
e fresa redondeadora
i fresa per attondare
n afrondingsfrees
d Abrundfräser

1986 rounding file
[far]
f lime pour arrondir
e lima redondeadora
i lima per attondare
n afrondingsvijl
d Abrundfeile

1987 rounding-off
[saw]
f exécution d'arrondis
e redondeo
i arrotondamento
n rond bijwerken
d Ausführung von Abrundungen

1988 rounding-over bit; rounding-over cutter
[car/fre/woo]
f fraise à rayon
e fresa de redondear
i fresa per attondare
n afrondfrees
d Viertelstab(-Abrund)-Fräser

1989 rounding plane
[pla]
f mouchette
e cepillo redondeador
i pialla per arrotondare

n bolschaaf; rondschaaf
d Rundhobel

1990 **round-nosed chisel**
[chi/met/sto]
f ciseau à pic rond; ciseau à tailler les pattes d'araignée; dégorgeoir
e cincel de pico redondo; cincel para cortar patas de arañas
i scalpello tondo per tagliare la scanalatura per la lubrificazione
n oliegroefbeitel
d Schmiernutmeissel

1991 **round-nosed wood turning chisel**
[chi/woo]
f ciseau rond de tourneur en bois
e cincel redondo de tornero de madera
i scalpello tondo di tornitore in legno
n ronde houtdraaibeitel
d Rundnasendrehbeitel

1992 **round nose pliers**
[pli]
f pince à becs ronds
e alicate(s) de bocas redondas
i pinza a becchi tondi
n punttang; rondbektang; tang met ronde bekken
d Rundzange; Zange mit runden Backen

1993 **round plane; round shave**
[coo]
f grattoir rond; plane à genoux
e plana rodillera
i sarchiatoio curvo
n afstroopschaaf
d Fasschaber; Rundschaber

1994 **round punch**
[per]
f perçoir; poinçon
e punzón
i foratoio; punteruolo; punzone
n doorn; doorslag; drevel
d Durchschläger

1995 **round set hammer**
[ham]
f dégorgeoir; fer à encoche
e martillo taladro
i martello a palla
n balhamer
d Ballhammer

1996 **round sharpening stone**
[gri/sha]
f pierre à affûter ronde
e piedra de amolar redonda
i pietra da affilare rotonda
n ronde aanzetsteen
d Rundschleiffeile

* **round shave → 608, 1993**

1997 **router**
[fre]
f fraiseuse verticale; mortaiseuse
e fresadora vertical
i fresatrice verticale
n bovenfreesmachine; vertikale freesmachine
d Oberfräser; Senkrechtfräser

1998 **router plane**
[pla]
f guimbarde
e guimbarda; rebajador
i pialla per sagomare
n grondschaaf
d Grundhobel

* **router plane → 724, 1677**

1999 **rubber backing pad**
[cpt/gri/san]
f plateau caoutchouc
e platillo de goma
i piattorello a gomma
n rubbersteunschijf
d Gummischleifteller

2000 **rubber float**
[mas]
f taloche à plateau revêtu de mousse latex
e fratás forada de goma
i piattello d'intonaco gommato
n rubberschuurbord
d gummiertes Reibebrett

2001 **rubber mallet**
[ham]
f maillet à caoutchouc
e mazo de goma
i mazzuola in caucciù; mazzuola in gomma
n zachtrubberhamer
d Gummihammer

2002 **ruler**
[mea]
f règle
e regla
i riga
n liniaal
d Lineal

* **rung → 1976**

* **runner stick** *GB* **→ 1148**

* **running center(re) → 1534**

S

* **sabre saw → 175, 1837**

2003 **saddler's awl**
[lea/per]
f alène de sellier
e lezna redonda del guarnicionero
i lesina rotonda di sellaio
n gaatels; zadelmakerspriem
d Riemenpfriem; Sattlerahle

2004 **saddler's hammer**
[ham/lea]
f marteau de sellier
e martillo de guarnicionero
i martello di sellaio
n zadelmakershamer
d Sattlerhammer

* **saddler's hollow punch → 198**

2005 **safe edge**
[far]
f bord lisse
e canto liso
i bordo liscio
n spaarkant
d glatte Schmalseite

2006 **safety bolt**
[fas/iwa/saf]
f boulon de sécurité
e perno de seguridad
i bullone di sicurezza
n borgbout
d Scherbolzen

2007 **safety bow; safety brace**
[fas/iwa/saf]
f bride de sécurité
e estribo de seguridad
i staffa di sicurezza
n borgbeugel
d Scherbügel

* **safety lamp → 708**

2008 **sailmaker's awl**
[nav/per]
f aiguille à voiles
e lezna de velero; pasador de velero
i caviglia di velaio
n zeilpriem
d Segelpfriem

2009 **saixe; sax; slate axe**
[cov/sto]
f hache d'ouvrage de couvreur
e azuela para desbastar pizarras
i scure per lavorare l'ardesia
n leikloverhak
d Schieferhacke

2010 **sander-grinder**
[gri/san]
f ponceuse-meuleuse
e lijadora-esmeriladora
i levigatrice-smerigliatrice
n schuur- en slijpmachine
d Sander-Schleifmaschine

* **sander-polisher → 1818**

2011 **sanders**
[san]
f ponceuses
e lijadoras
i levigatrici
n slijpmachines
d Schleifmaschinen

* **sanding disk → 5**

* **sanding ring → 7**

* **sanding roll → 8**

2012 **sandpaper**
[cpt/san]
f feuille abrasive
e papel abrasivo; papel de lija
i carta vetrata; carta smeriglio
n vel schuurpapier
d Schleifblatt

2013 **sash brush; window brush**
[dec]
f pinceau incliné
e pincel inclinado
i pennello per telaio di finestra
n bokkepoot
d Fensterrahmenpinsel

2014 **sash mortise chisel**
[chi]
f bédane étroit
e cincel estrecho
i bedano stretto
n kantbeitel; smalle schietbeitel
d schmaler Lochbeitel

2015 **saucer-shaped wire brush**
[cpt/pol]
f brosse-soucoupe
e brocha de cazuela cónica
i spazzola a tazza conica

n schotelvormige borstel
d Tellerbürste

2016 saw
[saw]
f scie
e serrucho; sierra
i sega; seghetta
n zaag
d Säge

2017 saw arbor
[cpt/foy/saw]
f arbre de scie circulaire
e árbol de sierra circular
i albero di sega circolare
n cirkelzaagas
d Kreissägenachse

2018 saw blade
[cpt/saw]
f lame de scie
e hoja de sierra
i lama di sega
n zaagblad
d Sägeblatt

2019 saw blade with staggered teeth
[cpt/saw]
f lame de scie à denture à hélice alternée
e hoja de sierra de dientes alternados
i lama di sega a dentatura spostata
n zaagblad met verzette vertanding
d Sägeblatt mit versetzter Zahnung; Segment-Kaltsägeblatt

2020 sawbuck *US*; **sawhorse** *GB*
[cpt/saw]
f chevalet; chèvre
e burro; cabellete para serrar; tijera
i cavalletto per segare
n zaagbok
d Sägeblock

2021 saw chain
[cpt/saw]
f chaîne de scie
e cadena de sierra
i catena di sega
n zaagketting
d Sägekette

2022 sawdust
[saw]
f sciure
e aserrín
i segatura
n zaagsel
d Sägemehl

2023 saw file
[cpt/saw]
f lime à scies; tiers-point
e lima para sierras
i lima per seghe
n zaagvijl
d Sägefeile

2024 saw for inlaid work
[saw]
f scie à marquetterie
e sierra de marquetería
i sega per intarsiatura
n aderzaag; marquetteriezaag
d Laubsäge; Säge für Einlegearbeit

2025 saw frame
[cpt/saw]
f monture de scie
e bastidor de sierra
i telaio per sega
n zaagraam
d Sägegestell

2026 saw grip
[cpt/saw]
f poignée
e asidero
i impugnatura
n zaaggreep; zaaghecht
d Sägehandgriff

2027 saw guide
[cpt/saw]
f guide-chaîne; passant
e llanta de la cadena
i guida della sega
n kettinggeleider
d Sägeschiene

2028 saw guide
[cpt/saw]
f guide-ruban
e guía de la cinta
i guida del nastro
n lintgeleider
d Bandführung

* **sawhorse** *GB* → **2020**

2029 sawing
[saw]
f sciage
e aserradura
i segatura
n zagen
d Sägen

2030 sawing cutter
[cpt/fre/saw]
f scie fraiseuse
e sierra fresadora
i sega fresatrice
n zaagfrees
d Sägefräser

2031 sawing table
[saw]
f table de scieur
e mesa para serrar
i tavola da segatore
n zaagtafel
d Sägetisch

2032 sawing table tilt
[saw]
f inclinaison de la table de scieur
e inclinación de la mesa para serrar
i inclinazione della tavola da segatore
n helling van de zaagtafel
d Neigung des Sägetisches

2033 saw kerf
[saw]
f coupe de scie; trait de scie
e corte de sierra
i tratto di sega
n zaagsnede
d Sägeschnitt

2034 sawmill
[saw]
f scierie
e aserradero
i segheria
n zagerij
d Sägewerk

2035 saw pit
[saw]
f fosse de scieur de long
e foso de serrador al hilo
i fosso di segatore per il lungo
n zaagput
d Sägegrube

2036 saw set
[cpt/saw]
f pince à voie
e tenaza de triscar; triscador
i tenaglie per licciare
n zaagzettang
d Schränkzange

2037 saw set; teeth setting iron
[cpt/saw]
f tourne-à-gauche
e triscador
i licciaiola per denti da sega; stradasega
n zaagzetter
d Sägensetzer; Schränkeisen

2038 saw sharpener
[far/saw]
f monture à tiers-point pour scies
e afiladora de sierras; entenalla para limar dientes de sierra
i sagoma ad affilatrice per seghe
n zaagvijlmal
d Sägenschärfkluppe

2039 sawtooth
[cpt/saw]
f dent de scie
e diente de sierra
i dente di sega
n zaagtand
d Sägezahn

2040 sawyer
[saw]
f scieur
e aserrador; serrador
i segatore
n zager
d Säger

* **sax → 2009**

2041 scabbling hammer; scappling hammer
[ham/sto]
f marteau de smillage; smille
e escoda
i martello da sbozzatura
n bikhamer
d Pickhammer

* **scaffolding hammer → 935**

* **scaling hammer → 275**

2042 scarf; scarf joint
[car/joi/wool]
f assemblage à trait de Jupiter droit
e empalme endentado; ensambladura con resalte
i giunto a doppio dente con testa diritta; unione a denti
n rechte haaklas
d gerades Hakenblatt

2043 scraper
[car/wool]
f grattoir; racloir
e raspador
i raschietto

n schraper
d Schrapper

2044 scraper; scraping iron
[dec]
f grattoir; racloir
e raspador; roscador
i raschietto
n afkrabber; krabber; schilderskrabber
d Kratze; Schaber

2045 scraper holder
[cpt/dec]
f porte-grattoir
e portarroscador
i portaraschietto
n krabbersteel
d Schaberhalter

*** scraper plane → 757**

*** scraping iron → 2044**

2046 scraping knife
[dec]
f couteau-racloir; dague
e cuchillo rascador
i coltello-raschietto
n afschraapmes
d Abstreichmesser

2047 scratch
[gen/met/pla]
f égratignure; raie; rayure
e rascadura
i graffiatura
n kras
d Kratz

2048 scratch brush
[cpt/gri]
f brosse à frotter
e cepillo de fregar; cepillo de púas de acero
i spazzola di metallo per raschiare
n schuurborstel
d Scheuerbürste

2049 scratch brush with scraper; wire brush with scraper
[cle]
f brosse à frotter à grattoir
e cepillo de fregar con roscador
i spazzola di acciaio con raschietto
n schuurborstel met krabber
d Scheuerbürste mit Schaber

2050 scratcher
[mas]
f gratteau
e arañador
i graffietto
n kraspen; ritsijzer
d Kratzeisen

2051 screen
[cle/sep/sto]
f crible à gravier; sas
e criba de arena; tamiz de arena
i vaglio per sabbia o ghiaia
n grind- en zandzeef
d Sandsieb

2052 screenings
[cle/sep/sto]
f criblure; poussier
e desperdicios del cribado
i rimasuglio di crivellatura
n uitziftsel
d Siebabfall

2053 screw-auger
[dri]
f tarrière hélicoïdale
e barrena helicoidal; barrena salomónica
i mecchia elicoidale a tortiglione
n avegaar met boorspaanbaan
d Schneckenbohrer

2054 screw-cutting chisel; thread-cutting chisel
[chi/scr]
f ciseau à fileter
e escoplo de roscar
i scalpello per filettare
n draadsnijbeitel; schroefdraadfrees
d Gewindedrehmeissel; Gewindestahl

2055 screw-cutting lathe; thread-cutting lathe
[lat/scr]
f tour à fileter
e torno para roscar
i tornio per filettare
n draadsnijbank
d Leitspindeldrehmaschine

2056 screwdriver
[scr]
f tournevis
e destornillador
i cacciavite
n schroevedraaier
d Schraubendreher

2057 screwing
[gen/scr]
f vissage
e atornillamiento
i invitatura

n inschroeven
d Anziehen

2058 screwing die
[scr]
f filière
e terraja
i filiera
n draadsnijijzer
d Gewindeschneideisen

2059 screwing die for gas pipes
[cut/hyd]
f filière pour filets de gaz
e terraja para hilos de gas; terraja para tubos de gas
i filiera per tubi da gas
n gasdraadsnijijzer
d Gasgewindekluppe; Gasgewindeschneidkluppe

2060 screwing machine
[scr]
f machine à fileter
e máquina de roscar
i filettatrice
n draadsnijmachine
d Gewindeschneidmaschine

2061 screwing pattern
[scr]
f bobine de filetage
e bobina para roscar
i bobina da filettare
n draadsnijpatroon
d Gewindeschneidpatrone

2062 screwing tools
[scr]
f outils de filetage
e útiles para roscar
i arnesi da filettare
n draadsnijgereedschap
d Gewindeschneidwerkzeuge

2063 screw nail
[iwa/scr]
f clou fileté
e clavo roscado
i vite autofilettante
n schroefnagel
d Schraubnagel

*** screw nail → 819**

2064 screw plate
[met/scr]
f filière à truelle; filière simple
e placa para roscar
i filiera semplice
n snijplaat
d Schneidkluppe

2065 screw shackle; turnbuckle
[iwa]
f manchon à vis
e grillete con pasador roscado
i tenditore
n koppelbus met schroefdraad
d Schraubmuffe

2066 screw slotting cutter
[fre/scr]
f fraise à rainer les têtes de vis
e fresa para hender cabezas de tornillos; ranuradora para cabezas de tornillos
i fresa per il taglio della testa delle vite
n schroefsleuffrees
d Schraubenschlitzfräser

2067 screwthread
[scr]
f filetage
e hilo de rosca
i filettatura
n schroefdraad
d Schraubengang; Schraubengewinde

2068 screwthread file; threading file
[far/scr]
f lime de filetage
e lima de roscas
i lima di filettatura
n draadsnijvijl; schroefdraadvijl
d Gewindefeile

*** screw wrench → 22**

2069 scriber
[gen/met/scr]
f pointe à tracer
e punta de trazar
i punta a tracciare
n kraspen
d Reissnadel; Stahlgriffel

2070 scribing block; surface ga(u)ge
[meal]
f trusquin à équerre
e gramil de mármol de trazar
i truschino
n krasblok
d Anreissgerät

*** scribing gouge → 1711**

2071 scroll chuck; self-centring chuck; universal chuck

[dri/lat/tool]
f mandrin à centrage automatique
e mandril autocentrante; mandril con ajuste espiral
i mandrino autocentrante
n boorhouder met automatische centrering
d Bohrfutter mit selbsttätiger Zentrierung

* **scroll saw** *US* → **1105**

* **scroll-saw blade** *US* → **1106**

* **scroll-saw frame** *US* → **1107**

* **scrubber** → **662**

2072 sculptor's chisels
[art/chi/sto]
f ciseaux de sculpteur
e cinceles de escultor; escoplos de escultor
i scalpelli di scultore
n beeldhouwersbeitels *(steen)*
d Bildhauermeissel *(Stein)*

2073 sculptor's rasp
[art/far]
f râpe de sculpteur
e escofina de escultor
i raspa di scultore
n beeldhouwersrasp
d Bildhauerraspel

2074 scutch; scutch holder; scutch tool
[art/chi/sto]
f égratignoir
e estique
i gradina
n gradinebeitel; kareerijzer; schareerijzer; tandijzer
d Scharriereisen

* **scutch hammer** → **471**

2075 scythe
[agr]
f faux
e guadaña
i falce
n zeis
d Sense

2076 scythe blade
[agr/hor]
f lame de faux
e hoja de guadaña
i lama di falce
n zeisblad; zeismes
d Sensenblatt

2077 scythe bladed sickle
[agr/hor]
f fauchette à lame cintrée
e hoz de hoja de guadaña
i falcetto
n kleine sikkel
d Sensensichel

* **scythe edge** → **686**

* **scythe handle** → **1286, 1287**

2078 scythe heel
[agr/cpt/hor]
f talon de faux
e talón de guadaña
i sperone di falce
n zeisrug
d Sensenbart

2079 scythe sheath
[agr/cpt/hor]
f couvre-lame
e guarda de filo
i guaina di falce
n zeisschede
d Sensenschuh

2080 scythe stone
[agr/sha]
f pierre à faux
e afiladora de la guadaña
i pietra da cote; pietra di falce flenaia
n wetsteen
d Sensenschärfer

2081 sealing pliers
[pli/tra]
f pince à plomber
e alicate(s) para precintar; alicate(s) precintadora(s)
i pinza per impiombare
n plombeertang
d Plombenzange; Plombierzange

* **seam drill** → **1791**

* **secateur** → **1857**

2082 secondary table
[cpt/dri]
f surtable
e tabla superior
i piano superiore
n bovenblad
d Übertisch

2083 second cut
[far]

f taille demidouce
e entrefina; picadura semidulce
i taglio mezzodolce
n halfzoete kap
d Hieb 2

* **second tap → 1440**

* **secret dovetail → 1620**

* **secret dovetail joint → 764**

2084 **sector garden sprinkler**
[hor]
f arroseur de secteur
e aspersor sectorial
i irrigatore di settore
n sectorsproeier
d Sektorsprenger

2085 **security screw**
[saf/scr]
f vis d'arrêt; vis de serrage
e tornillo de fijación
i vite di fissaggio
n borgschroef
d Hemmschraube; Klemmschraube; Sicherungsschraube

2086 **Seeger ring pliers**
[pli]
f pince pour anneaux Seeger
e alicate(s) para arandelas de Seeger
i pinza per anelli Seeger
n seegerringtang
d Seegerringzange

2087 **self-adjusting level; self-level(l)ing level**
[mea/sur]
f niveau à autonivellement; niveau automatique
e nivel de autonivelación
i livella ad autolivellazione
n waterpas met zelfnivellering
d automatisch horizontierende Libelle; Libelle mit Selbstnivellierung

2088 **self-cent(e)ring punch**
[per]
f poinçon à centrage automatique
e punzón autocentrante
i punzone autocentrante
n automatische centerpons
d selbsttätiger Zentrierkörner

* **self-centring chuck → 2071**

2089 **self-drilling anchor**
[cpt/dri/fas]
f cheville autoforeuse
e taco autoperforador
i perno autoperforatore
n zelfborend anker
d Selbstbohrdübel

* **self-emptying bit → 1408**

* **self-grip wrench → 1624**

* **self-level(l)ing level → 2087**

2090 **self-lubricating bearing**
[eng/lub]
f palier autolubrifiant
e cojinete autolubricante
i cuscinetto autolubrificante
n zelfsmerend lager
d selbstschmierendes Lager

2091 **self-setting clamp**
[cla/gen]
f serre à blocage automatique
e mordaza autofraguada
i morsa a bloccaggio automatico
n zelfsluitende klem
d selbstschliessende Klemme

2092 **self-tapping screw**
[scr]
f vis à autofiletage
e tornillo de autorroscado
i vite autofilettante; vite maschiatrice
n zelftappende staalschroef
d selbstschneidende Schraube

* **self-tightening chuck → 100**

2093 **self-winding cable reels**
[ele/hor/wel]
f prolongateur-enrouleur à enroulement automatique
e tambor de cable de arrollamiento automático
i avvolgicavo a avvoltato automatico
n kabeltrommel met automatische oprolling
d Kabeltrommel mit Rückhol-Automatik

2094 **semi-automatic spade**
[agr/hor]
f bêche semiautomatique
e laya semiautomática
i vanga semiautomatica
n halfautomatische schop
d halbautomatischer Spaten

2095 **semi-circular cutting edge**
[chi/cpt]
f biseau de coupe semicirculaire

e bisel de corte semirredondo
i smusso di taglio semicircolare
n halfronde snede
d halbrunde Fase

2096 semi-flexible grinding disk
[cpt/gri]
f disque à meuler semirigide
e muela semiflexible
i mola abrasiva semiflessibile
n halfstijve slijpschijf
d halbsteife Schleifscheibe

* **serrated-edge trowel → 14**

2097 service ell
[hyd]
f raccord coudé en L à filetage intérieur et extérieur
e codo en L con una rosca exterior y otra interior
i gomito con una estremità filettata esternamente ed altra internamente
n kniestuk met een inwendige en een uitwendige schroefdraad
d Kniestück mit innerem und äusserem Gewinde

2098 set
[saw]
f avoyé
e triscado
i allicciato
n gezet
d geschränkt

2099 set screw
[cpt/scr]
f vis d'arrêt; vis de réglage
e tornillo de ajuste
i vite di fermo; vite di pressione
n stelschroef
d Stellschraube

2100 setting
[saw]
f avoyage
e triscado
i allicciatura
n zetten
d Schränkung

2101 sextant
[mea/nav]
f sextant
e sextante
i sestante
n sextant
d Sechstelkreis; Sextant

* **shaft → 1281**

2102 shank *(drilling bit)*
[cpt/dri/gen]
f queue; tige
e caña; espiga
i stelo
n schacht; schaft
d Schaft

2103 shank *(rivet)*
[riv]
f branche; tige
e caña
i asta
n pen; stift
d Schaft

2104 shank *(screw)*
[scr]
f corps
e caña; espárrago; espiga
i gamba
n steel
d Schaft

2105 shank *(screwdriver)*
[scd]
f lame de tournevis
e hoja de destornillador
i lama di cacciavite
n kling van een schroevedraaier
d Klinge eines Schraubendrehers

2106 shaper tools
[tool]
f outils de mortaisage
e herramientas de mortajar
i utensileria per stozzare
n stootbeitels
d Stossmeissel; Stosstähle

2107 shaping hammer
[ham/met]
f marteau à décolleter; marteau à façonner; marteau à profiler
e martillo de joyero
i martello per profilatura
n façonneerhamer
d Fassonhammer

2108 shaping machine
[mec/tool]
f mortaiseuse horizontale
e limadora; mortajadora horizontal
i macchina stozzatrice orizzontale
n horizontale stootmachine
d Waagerecht-Stossmaschine

2109 sharpening
[gri/sha]
f affilage; affûtage; aiguisage; émoulage
e afilado; amolado
i affilatura
n scherpen; slijpen; wetten
d Schärfen; Schleifen

2110 sharpening disk
[gri/sha]
f roue d'appointissage
e disco para aguzar
i disco da appuntatura
n aanpuntschijf
d Anspitzscheibe

2111 sharpening file
[far/sha]
f affiloir; lime d'affûtage
e lima afiladora; lima vaciadora
i lima da affilatura
n slijpvijl
d Schärffeile; Schleiffeile

*** sharpening steel → 1210, 1481**

2112 sharpening stones
[gri/sha]
f pierres à affûter
e piedras de amolar
i pietre da affilare
n aanzetstenen
d Banksteine; Schleifsteine

2113 shave hook
[dec]
f grattoir-raclette; racloir
e rascador
i raschiatoio
n verfkrabber
d Kratzeisen

*** shave hook → 1689**

*** shaving horse → 605**

2114 shavings
[pla]
f copeaux
e acepilladura; virutas
i trucioli
n krullen
d Hobelspäne

2115 shear legs; sheer legs; sheers
[nav/tra]
f bigue; chèvre à haubans
e cabria; cabria de tres pies
i capra a tre piedi
n bokkebeen; schaarmastkraan
d Scherenkran

*** sheave → 1162**

2116 sheet metal shears
[she]
f cisailles à tôles
e cizallas de chapa
i cesoie di lamiera
n blikscharen
d Blechscheren

*** sheet planishing hammer → 1783**

2117 sheet saw
[saw]
f scie à tôle
e sierra para chapas
i sega per lamiere
n blikzaag
d Blechsäge

*** sheet shears** *GB* **→ 1784**

*** shell bit → 1182**

2118 shell countersink
[dri/rea]
f alésoir-fraise creux
e avellanador postizo
i accecatoio cavo
n opsteekverzinkfrees
d Aufstecksenker

2119 shell gimlet
[dri]
f vrille à cuiller
e barrena de cuchara
i succhiello a cucchiaio
n lepelfretboor
d Löffelbohrer

2120 shell reamer
[dri/fre/rea]
f alésoir mécanique creux
e escariador mecánico hueco; escariador mecánico postizo
i alesatore meccanico cavo
n opsteekmachineruimer; opsteekruimer
d Maschinenaufsteckreibahle

2121 shepherd's staff
[agr]
f houlette
e cayado pastoril; garrota
i verga del pastore
n haakschop; herdersschop
d Hirtenstab

2122 shim
[iwa]
f cale de support
e chapa de relleno
i spessore
n sluitplaatje; steunplaatje
d Ausgleichscheibe; Unterlegstück

* **shingle rake → 1200**

* **shingle roller → 1201**

2123 shipbuilder's chisel
[chi/nav]
f fermoir pour constructeurs de vaisseaux
e formón de constructor de buques
i scalpello per costruttori navali
n kaafijzer; kapijzer; scheepsbeitel
d Schiffbauerbeitel

* **shoeing hammer → 938**

* **shoe knife → 550**

* **shoe rasp → 551**

2124 shoe stretcher
[lea]
f machine à élargir chaussures
e máquina de ahormar zapatos
i forma per allargare scarpe
n schoenenrekapparaat
d Schuhausweitmaschine

2125 shook
[coo]
f botte de douves
e paquete de duelas
i fascio di doge
n duigen voor een vat
d Bündelfassdauben

2126 shooting
[car/joi/woo]
f dressage d'une planche
e canteado de una tabla
i piallatura dei bordi di un'asse
n afschaven van planken
d Abhobeln von Brettern

* **shooting plane → 1466**

2127 shop knife; trimming knife; universal knife
[cut/gen]
f couteau du bricoleur; couteau universel
e cuchillo de bricolador
i coltello di uso universale
n karweitjesmes; vakmes
d Hobbymesser; Vielzweckmesser

2128 short crowbar
[bui/eng/gen]
f petit pied de biche
e palanqueta
i piccolo pie di porco
n handijzer
d kurze Brechstange; kurzes Brecheisen

2129 short snipe nose pliers
[pli]
f pince à becs courts affilés
e alicate(s) de bocas planas cortas y puntiagudas
i pinza a becchi piatti corti appuntati
n tang met korte platte spits uitlopende bekken
d Kurzbeckzange mit spitz zulaufenden flachen Backen

2130 shot drill
[dri/min/sto]
f trépan de choc
e barrena de percusión; barrena percutora
i trivella a percussione
n slagboor; stootboor
d Schlagbohrer; Stossbohrer

2131 shot drilling
[dri/min/sto]
f trépanage de choc
e barrenado de percusión
i trivellatura a percussione
n slagboren; stootboren
d Schlagbohren; Stossbohren

2132 shoulder of a chisel
[chi/cpt]
f épaulement de ciseau
e espaldón de cincel
i spallamento di scalpello
n borst van een beitel
d Beitelschulter

* **shoulder plane → 1914**

2133 shoulders
[car/joi/woo]
f épaulements
e muñones
i tenoni
n borsten
d Schulter; Zapfen

2134 shoulder saw
[saw]
f scie d'épaulement

e sierra de muñón
i sega di tenone
n afzetzaag; borstzaag; boutzaag
d Schultersäge; Zapfensäge

* **shuttering hammer → 1092**

* **shuttering plane → 1093**

* **sickle → 1911**

* **sickle spanner → 1397, 1750**

2135 side-and-face cutter
[fre]
f fraise à surfacer en bout; fraise cylindrique en bout
e fresa cilíndrica para refrentar; fresa de tres cortes
i fresa laterale da spianare
n mantelkopfrees
d Walzenstirnfräser

2136 side cutter
[fre]
f fraise latérale
e fresa de dientes laterales; fresa de disco
i fresa laterale
n schijffrees
d Scheibenfräser

2137 side cutting pliers
[pli]
f pince coupante de côté
e alicate(s) de corte oblicuo
i pinza a taglio laterale
n zijkniptang; zijsnijtang
d Seitenschneider

2138 side handle
[dri/gen]
f poignée latérale
e empuñadura lateral
i impugnatura laterale
n zijhandgreep
d seitlicher Griff

2139 side rail
[cpt/fir/saf]
f montant d'échelle
e larguero de la escalera
i staggio di scala
n boom van een ladder
d Holm eines Leiters

2140 side-rebate plane
[pla]
f guillaume latéral
e guillame lateral
i sponderuola laterale
n overzijschaaf
d Seitenfalzhobel

2141 side tool
[fre]
f outil couteau
e cuchillo acodillado
i utensile laterale per tornio
n mesbeitel; zijbeitel
d gebogener Seitenstahl; Seitendrehmeissel

* **side tool → 1483**

2142 sieve
[gen/sep]
f crible; tamis
e criba; tamiz
i setaccio; vaglio
n zeef
d Sieb

* **sight rule → 39**

* **silence dome → 748**

* **silicon carbon wheel → 407**

2143 simple press tool
[too]
f matrice simple pour presses
e estampa sencilla; traquel sencilla de prensa
i matrice semplice di pressa
n eenvoudige persmatrijs
d einfache Pressmatrize

2144 sine bar
[mea]
f goniomètre de précision
e barra de senos; goniómetro de precisión
i barra seno
n precisiegoniometer
d Präzisionsgoniometer

2145 single angle cutter
[car/fre/woo]
f fraise angulaire d'un côté
e fresa angular de un labio
i fresa angolare unilaterale
n frees met enkele snijkant
d einseitiger Winkelfräser

2146 single bit drilling
[dri]
f perceuse à broche simple
e taladro monohusillo
i trapano a mandrino solo

n boormachine met één spil
d Einspindelbohrmaschine

2147 single cut
[far]
f première taille; simple taille
e picadura de fondo
i taglio semplice
n enkele kap
d Grundhieb

2148 single flute bit; single flute cutter
[car/fre/wool]
f fraise droite à une cannelure
e fresa de un corte
i fresa ad uno taglio
n eenzijdige frees
d einseitiger Fräser; Einzelschneider

2149 single-spindle automatic
[lat/scr]
f tour automatique à une broche
e torno automático monohusillo
i tornio ad una punta; tornio ad uno mandrino
n eenspillige automaat
d einspindeliger Automat

2150 skew-back blade
[cpt/saw]
f lame à dos aminci
e hoja con costilla hueca
i lama con dorso cavo
n zaagblad met gegolfde rug; zaagblad met holle rug
d Sägeblatt mit geschweiftem Rücken

2151 skew bull-nose plane
[car/pla]
f rabot à fer oblique
e cepillo de hierro oblicuo
i pialla a ferro obliquo
n schaaf met schuinlopende beitel
d Schräghobel

2152 skew nailing
[car/iwa]
f clouage incliné
e acción de clavar inclinado
i chiodatura obliqua
n schuin spijkeren
d Schrägnageln

2153 skew rebate plane
[pla]
f guillaume oblique
e guillame oblicuo
i sponderuola obliqua
n schuine rabatschaaf
d schräger Falzhobel

2154 skimmer float; wooden float
[mas]
f taloche à bois
e fratás de madera
i piattello di legno
n schuurbord *(hout)*
d Reibebrett *(Holz)*

2155 skip-tooth saw
[saw]
f scie à dents distanciées
e sierra de dientes distanciados
i sega a denti distanziati
n zaag met grote ruimte tussen de tanden
d Säge mit grossem Raum zwischen den Zähnen

2156 slabber; trimming saw
[fre/saw]
f machine à fraiser genre raboteuse; scie à corroyage du bois
e fresadora-cepillo
i fresatrice pialla
n freesschaaf
d Langfräsmaschine

2157 slabbing
[car/wool]
f sciage de troncs d'arbres en poutres carrées
e labra de troncos en vigas cuadradas
i squadratura dei tronchi
n in vierkante balken verzagen van boomstammen
d Baumstämme zersägen in rechteckigen Balken

*** slate axe → 2009**

2158 slater's hammer
[cov/ham]
f marteau d'ardoisier
e martillo de pizarrero
i martello di copritetti di lavagna
n leidekkershamer
d Schieferdeckerhammer

2159 slaughtering hammer; slaughtering mallet
[ham/nut]
f merlin
e mazo de matanza
i mazzapicchio
n dolhamer; kolhamer
d Schlachthammer

2160 slaughtering mask
[nut]
f masque Bruneau
e máscara de matanza
i maschera da tiro
n slachtmasker
d Schlachtmaske

* **slaughterman's tools** *GB* → **993**

2161 sledge; sledge hammer
[ham]
f maillet en bois
e maceta de madera
i mazzuola a legno
n sleg(ge); slegel; slei
d Schlegel

* **sledge → 1**

* **sledge hammer → 1**

2162 sleekers; slickers; smoothers
[fou]
f lissoirs
e alisadores
i ferri di angolo; lisciatoio; spatole a lisciare
n polijstnokken; slikkers
d Polierknöpfe

2163 sleeve
[fas]
f manchon
e manguito
i bussola; manicotto
n bus; manchet; mof
d Büchse; Hülse; Muffe

2164 sleeve piece; thimble
[hyd]
f bague d'assemblage; virole
e manguito de acoplamiento
i manicotto di giunzione
n koppelmof
d Muffenkupplung

2165 slick
[chi]
f lissoir
e formón ancho de alisar
i scalpello largo per lisciare
n brede polijstbeitel
d Glätteisen

2166 slicker; slicking stick
[lea]
f astic; guinche
e bisagra; costa
i bisegolo; bussetto
n gladhout; likbeen
d Glättholz

2167 slicker
[lea/pol]
f étire
e alisador de pieles
i brunitoio; scarnatoio
n schaafijzer
d Schabeisen

* **slicker → 335**

* **slickers → 2162**

* **slicking stick → 2166**

2168 slide cal(l)ipers; slide ga(u)ge; vernier cal(l)ipers
[meа]
f pied à coulisse
e calibrador de cursor; pie de rey
i calibro a corsoio
n schuifmaat
d Schieblehre

2169 slide rest
[lat/too]
f chariot transversal
e carro transversal
i carrello trasversale
n dwarsslede
d Planschlitten; Querschlitten

2170 slide rule; sliding rule
[mea]
f règle à calcul
e regla de cálculo
i regolo calcolatore
n rekenliniaal; rekenschuif
d Rechenschieber

2171 sliding bevel
[car/mea/woo]
f fausse équerre
e falsa escuadra; sentanilla
i squadra a lati mobili; squadra falsa
n zwaaihaak; zwei; zweihaak
d Schmiege; Schrägmass; Stellwinkel; Winkelpasser

* **sliding rule → 2170**

2172 slip ga(u)ge
[mea]
f calibre d'épaisseur
e calibrador de espesores
i calibro per spessori; calibro piatto di

precisione
n diktekaliber
d Fühlerlehre

2173 slip joint pliers
[pli]
f pince universelle réglable à deux positions
e alicate(s) de dos posiciones; alicate(s) extensible(s)
i pinza a cerniera snodata
n pijptang; tang met verstelbare bekopening
d in zwei Positionen verstellbare Kombizange

* **slip stone → 1184**

* **slit of a plane → 1635**

2174 slitting file
[far]
f lime fendante
e lima achaflanada; lima hendedora
i lima a mandorla
n sleufvijl
d Nutenfeile

* **slitting saw → 1068**

2175 slogging ring spanner
[spa]
f clé polygonale à frapper
e llave golpe boca abierta
i chiave poligonale da battere
n ringslagsleutel
d Ringschlagschlüssel

2176 sloped toothing
[cpt/saw]
f denture couchée
e dentadura angular; dentadura oblicua
i dentatura obliqua
n schuine vertanding
d NV-Zahnung; Schrägverzahnung

2177 slope level
[sur]
f clinomètre
e clinómetro; eclímetro; nivel de pendientes
i clinometro
n clinometer; hellingmeter
d Neigungsmesser

* **slot cutter → 1217**

* **slot milling tool → 1217**

2178 slotting cutter
[fre]
f fraise à rainurer
e fresa para ranurar
i fresa per scanalare
n groeffrees
d Nutenfräser

2179 slotting tool
[chi]
f outil à mortaiser
e útil de mortajar
i utensile per mortasatura
n slotgatgereedschap
d Ausklinkwerkzeug; Stossstahl

* **smith's mate → 1265**

* **smithy → 1085**

2180 smoke annihilator
[cle/saf]
f fumivore
e guardahumos
i parafumo
n rookvang
d Rauchverzehrer

2181 smooth
[far]
f doux
e dulce
i dolce
n zoet
d schlicht

2182 smooth cut
[far]
f taille douce
e picadura dulce
i taglio dolce
n zoete kap
d Hieb 3

2183 smooth-cut file
[far]
f lime douce
e lima dulce
i lima dolce
n gladvijl; zoetvijl
d Abziehfeile; Hieb-3-Feile; Schlichtfeile

* **smoothers → 2162**

* **smooth finishing → 776**

2184 smooth grinding; smoothing
[gri/pol]
f adoucissage à la meule; lissage

e alisadura
i lisciatura
n gladslijpen
d Glattschleifen

2185 smoothing plane
[pla]
f rabot plat; varlope à recaler
e cepillo acabador; cepillo de alisar
i pialla da rifinire
n gladschaaf; kleine blokschaaf; zoetschaaf
d Schlichthobel

2186 smooth-sided flat file
[far]
f lime plate à deux bords lisses
e lima plana de dos cantos lisos
i lima piatta a due bordi lisci
n platte vijl met gladde kanten
d Flachfeile mit glatten Schmalseiten

2187 snap; snap-tool
[iwa/met]
f bouterolle
e buterola
i stampa per ribattini
n dopbeitel; dopper; snapper
d Döpper; Nietkopfmacher

2188 snap ga(u)ge
[mea]
f calibre à fers de cheval; mâchoire double
e calibre de mordazas
i calibro a gole; calibro a forcella; calibro a forchetto
n bakkaliber; grenskaliber
d Grenzlehre; Rachenlehre

2189 snap-head
[iwa/met]
f tête bouterollée
e cabeza redonda de remache
i testa tonda di ribattino
n ronde klinknagelkop
d runder Nietkopf

*** snap-head rivet → 1984**

*** snap-tool → 2187**

2190 snipe nose pliers
[pli]
f pince à becs affilés
e alicate(s) de bocas puntiagudas
i pinza a becchi appuntati
n tang met spits uitlopende bekken
d Zange mit spitz zulaufenden Backen

2191 socket
[hyd]
f emboîtement; manchon de tuyau
e casquillo
i bussola
n fitting
d Röhrensockel

2192 socket chisel
[chi]
f ciseau à douille
e cincel de casquillo; escoplo de espiga hueca
i scalpello per mortasa
n lepelvormige schietbeitel
d Düllbeitel

2193 socket head screw
[iwa]
f vis à tête cylindrique à six pans creuse
e tornillo de cabeza cilíndrica con exágono interior
i vite ad incassatura esagonale
n binnenzeskantbout
d Zylinderkopfschraube mit Innensechskant

2194 socket spanner; socket wrench
[spa]
f clé à douille; clé à pipe
e llave de vaso
i chiave a bussola
n dopsleutel
d Hülsenschlüssel; Steckschlüssel

*** soft faced hammer → 1909**

*** soft faced mallet → 1909**

*** soil pulverizer** *US* **→ 1312**

2195 soldering blowlamp
[wel]
f pistolet à souder
e pistola de soldar
i cannello ferruminatorio
n soldeerpistool
d Lötpistole

2196 soldering copper; soldering iron
[wel]
f fer à souder; soudoir
e soldador
i saldatoio
n soldeerbout
d Lötkolben

2197 soldering disk
[wel]

f disque à souder
e plato de la fragua
i disco del fornello per saldare
n soldeerschijf
d Lötscheibe

2198 **soldering gun**
[wel]
f pistolet à souder
e pistola de soldar
i pistola per saldare
n soldeerpistool; soldeerrevolver
d Lötpistole; Schweisspistole

* **soldering iron → 2196**

2199 **soldering tongs**
[pli/wel]
f pince à souder
e alicate(s) para soldar
i pinza per saldare
n soldeertang
d Lötzange

2200 **sole of a plane**
[cpt/pla]
f semelle de rabot
e base de cepillo
i base di pialla
n schaafbaan; schaafzool
d Hobelsohle

* **sounder → 721**

2201 **spacer**
[cpt/gen/gri/san]
f rondelle d'écartement
e arandela separadora
i rondella distanziatrice
n sluitring
d Zwischenring

2202 **spade**
[agr/gen/hor]
f bêche
e pala
i badile; pala
n schop; spade
d Schaufel; Spaten

2203 **spalling**
[chi/sto]
f smillage
e desbaste de cantera
i scagliatura di martellamento
n steensplinters
d Absplitterung

2204 **spanners**
[spa]
f clés
e llaves
i chiavi
n sleutels
d Schlüssel

* **span saw → 1102**

2205 **spare grinding wheel**
[cpt/gri]
f meule de rechange
e muela de repuesto; muela de reserva
i mola di ricambio
n reserveslijpschijf
d Ersatzschleifscheibe

2206 **sparking plug brush** *GB*; **spark plug brush** *US*
[ele/tra]
f brosse pour bougies
e brocha para bujías; bruza para bujías
i spazzola per candele di accensione
n bougieborstel
d Zündkerzenbürste

2207 **sparking plug spanner** *GB*; **spark plug spanner** *US*
[ele/spa/tra]
f clé à bougies
e llave de estrella para bujías
i chiave per candele
n bougiesleutel
d Zündkerzenschlüssel

* **spark plug brush** *US* **→ 2206**

* **spear point chisel → 731**

2208 **speed reducer**
[cpt/dri]
f réducteur mécanique de vitesse
e contramarcha reductora
i riduttore meccanico di velocità
n mechanische toerenregelaar
d mechanisches Reduziergetriebe

2209 **spherical cutter**
[fre/gas]
f fraise sphérique
e fresa esférica
i fresa a sfera
n kogelfrees
d Kugelfräser

2210 **spherical stake**
[for]
f bigorne bombée

e bigornia esférica
i bicornia sferica
n bolstaak
d hochrunde Faust

* **spider spanner → 403**

* **spider wrench → 403**

2211 **spigot**
[hyd/nut]
f cheville; goupille
e espita
i colletto di centraggio
n stift
d Stift

2212 **spigot-and-socket joint**
[hyd/joi]
f joint à manchon et cheville
e junta de enchufe y casquillo
i giunto a bicchiere
n fitting-stift-verbinding
d Fassung-Stift-Verbindung

2213 **spike**
[iwa]
f clou à large tête
e clavo grueso
i chiodo grosso
n lange spijker
d langer Nagel

2214 **spile awl**
[dri]
f épitoir; évidoir; mèche creuse
e barrena hueca; broca hueca
i trapano cavo
n holle boor
d Hohlbohrer

2215 **spindle**
[cpt/gen]
f arbre; axe
e árbol; eje
i alberino; mandrino
n as; spil
d Achse; Spindel; Welle

2216 **spindle travel**
[cpt/dri]
f course de la broche
e carrera del vástago
i corsa del mandrino
n spilverplaatsing
d Spindelhub

2217 **spinning**
[lat/met]
f emboutissage au tour; repoussage
e embutido al torno; repulsado
i imbutitura al tornio
n kouddrukken
d Druckbandverfahren; Kaltdrücken

2218 **spiral-fluted reamer**
[dri/fre/rea]
f alésoir à cannelure hélicoïdale
e escariador de estrías helicoidales
i alesatore a scanalatura elicoidale
n ruimer met spiraalgroeven
d spiralgenutete Reibahle

2219 **spiral gear**
[mec/tra]
f engrenage à hélicoïde; engrenage à roues hélicoïdales
e engranaje de dentadura espiral
i ingranaggio elicoidale
n schroefradoverbrenging
d Schraubengetriebe; Schraubenradgetriebe

* **spiral ratchet screwdriver → 107**

2220 **spiral spring**
[cpt/gen]
f ressort à boudin
e muelle helicoidal; resorte en espiral
i molla spirale
n spiraalveer
d Schraubfeder; Spiralfeder; Sprungfeder

* **spirit level → 1522**

2221 **spit of a spade**
[agr/hor]
f profondeur de fer de bêche
e profundidad de la lama de pala
i profondità del ferro di badile
n steek van een schop
d Spatenstich

2222 **split bearing**
[civ/eng]
f coussinet à coquilles
e cojinete seccional
i cuscinetto diviso; cuscinetto in due pezzi
n tweedelig lager
d Schalenlager

2223 **split pin**
[iwa]
f goupille fendue
e grupilla; parador
i copiglia
n splitpen
d geschlitzter Stift; Splint

2224 split pulley
[eng/tra]
f poulie à deux pièces
e polea en dos partes
i puleggia spaccata
n tweedelige katrol
d geteilte Riemenscheiebe

* **splitting knife → 535**

* **spoke → 1976**

2225 spokeshave
[pla]
f vastringue; wastringue
e bastren
i pialletto per sagomare
n spookschaaf
d Schabhobel

* **spoon bit → 767**

2226 spoon bit chisel
[chi/woo]
f ciseau-cuillère
e cincel de cuchara
i scalpello a cucchiaio
n lepelbeitel
d Löffelbeitel

* **spoon bit gouge → 1115**

2227 spoon-shaped rasp
[far/lea]
f râpe en cuiller
e escofina en forma de cuchara; escofina gruesa
i raspa a cucchiaio
n grove rasp
d Löffelraspel

* **spoon tool → 857**

2228 spot board
[mas]
f taloche de maçon
e tabla cuadrada para el yeso amasado
i sparviere
n kalkbord; spaarbord
d Streichbrett

* **spotting drill → 448**

2229 spot welding pliers
[pli/wel]
f pince à soudage par points
e alicate(s) para soldadura de puntas
i pinza per saldatura a punti
n puntlastang
d Punktschweisszange

2230 spray gun
[dec/gen]
f pistolet de vernissage; pistolet pour peinture pneumatique
e pistola rociadora
i pistola a spruzzo
n verfspuit
d Farbspritzpistole

* **spray gun → 31**

2231 spray nozzle
[hor]
f lance d'arrosage
e lanza de regadera
i lance a getto
n straalpijp
d Spritzdüse

2232 sprig
[fas/iwa]
f pointe sans tête
e clavo sin cabeza; punta
i chiodo senza testa; punta
n spijkertje zonder kop
d Nagel ohne Kopf

2233 sprig
[fas/iwa]
f pointe
e alfiler de espiga
i asticina
n bagetnaald
d Stift

* **sprig bit → 318**

* **sprigging → 944**

2234 spring
[cpt/gen]
f ressort
e muelle
i molla
n veer
d Feder

2235 spring bows
[mea]
f balustre à ressort
e compás de muelle
i balaustrino
n veerpasser
d Federzirkel; Nullenzirkel

2236 spring cramp
[car/joi/wool]
f serre-joint à collage rapide
e sargento para colar
i pinze per incollare
n snellijmtang
d Schnelleimzange

* **spring cramp → 1902**

* **spring pawl → 1719**

2237 spring steel blade
[cpt/met]
f lame en acier à ressorts
e hoja de acero para muelles
i lama in acciaio per molle
n verenstaalblad
d Federstahlblatt

2238 sprinkling can *US*; **watering can** *GB*
[hor]
f arrosoir de jardin
e regadera
i annaffiatoio
n gieter
d Giesskanne

* **sprue stick** *US* **→ 1148**

* **spur → 1651**

* **spur wheel → 1748**

* **square file → 1873**

2239 square horn
[for]
f corne carrée
e cuerno escuadro
i corno quadro
n vierkante hoorn
d Rechteckhorn

* **square mallet → 415**

* **square nosed pliers → 825**

2240 square reamer
[dri/fre/rea]
f alésoir à quatre pans
e escariador de macho cuadrado; escariador de pitón cuadrado
i alesatore ad angoli retti
n rechthoekige ruimer
d Vierkantaufreiber

2241 square rebate plane
[pla]
f guillaume carré
e guillame cuadrado
i sponderuola quadra
n vierkante rabatschaaf
d rechteckiger Falzhobel

2242 square sharpening stone
[gri/sha]
f pierre à affûter carrée
e piedra de amolar cuadrada
i pietra da affilare quadra
n vierkante aanzetsteen
d Vierkantschleiffeile

2243 square thread
[iwa/scr]
f filetage à pas carré
e rosca cuadrada
i filettatura quadra
n rechthoekige schroefdraad
d rechtwinkliges Gewinde

2244 squaring edges
[san]
f mise à l'équerre des chants
e escuadro de cantos
i messa al quadrato di bordi
n haaks schuren van randen
d Winkelrechtfügen von Kanten; Winklig-schleifen von Kanten

2245 squaring shears
[met/she]
f cisailles à ébouter; cisailles à étêter
e tijeras de escuadrar
i cesoie a squadrare
n kopschaar
d Kopfschere

2246 stadia rod; subtense bar
[mea/sur]
f mire parlante; stadia
e barra estadimétrica; estadia; mira taquimétrica
i barra stadimetrica; stadia
n basislat
d Basismesslatte

2247 stake
[cpt/for]
f bigorne; enclumette
e bigorneta; bigornia
i bicornia; incudinetto
n staak
d Faust

* **stand for portable cut-off work → 674**

2248 staple
[fas/gen/iwa/off]
f agrafe
e grapa
i grappetta
n hechtnietje
d Heftklammer

2249 staple
[iwa]
f agrafe; crampe
e grampillón
i arpione; grappa
n kram
d Krampe

* **staple → 644**

2250 staple gun
[fas/gen/iwa/off]
f agrafeuse automatique
e pistola clava-grapas; pistola de engrapar
i pistola per grappette
n nietautomaat
d Heftklammerpistole

2251 stapler
[fas/iwa/gen/off]
f agrafeuse
e cosepapeles
i cucitrice con grappette
n nietapparaat
d Heftapparat

2252 stave
[coo]
f douve
e duela
i doga
n duig
d Daube

2253 stave jointing bank
[coo]
f banc pour le façonnage des douves
e paloma de tonelero
i pialla fissa per la rifilatura delle doghe
n sneepbank
d Fügebank; Stossbank

2254 stave template
[coo]
f gabarit de douve; panneau
e plantilla de duela
i sagoma di doga
n duigmal
d Daubenschablone

2255 stave wood
[coo]
f bois de douve
e madera de duela
i legno di doga
n duighout
d Daubenholz

2256 steady pin
[fas]
f cheville de fixation; étoquiau
e clavija de fijación; pasador de fijación
i chiavetta per fissare
n bevestigingspen; rijnagel; slotstift; stijpel
d Fixierstift

* **steel → 1481**

2257 steel brush; wire brush
[cle]
f brosse métallique
e cepillo de alambre
i spazzola metallica
n draadborstel
d Drahtbürste

2258 steel marking ga(u)ge
[mea]
f trusquin en acier
e gramil de acero
i graffietto di acciaio
n kruisstaal
d stählernes Streichmass

2259 steel nail
[fas/iwa]
f clou d'acier
e clavo de acero
i chiodo di acciaio
n staalnagel
d Stahlnagel

* **steel nail for concrete → 582**

2260 steel rule; straight edge
[mea]
f règle en acier
e regla de acero
i regolo di acciaio; riga di acciaio
n stalen lineaal; stalen meetlat; stalen rij
d Massstab aus Federstahl

2261 steel set square
[mea]
f équerre en acier
e escuadra de acero
i squadra di acciaio
n stalen winkelhaak
d Stahlwinkel

2262 steel wire sweeping broom
[cle/hou]
f balai en fils d'acier
e escoba de alambre de acero
i granata a fili di acciaio
n staaldraadbezem
d Stahldrahtbesen

* **steering hoe → 1370**

2263 stencil brush; stencilling brush
[art/dec]
f brosse pour patronner
e brocha para plantillas
i pennello per mascherina
n sjabloneerkwast
d Schablonenbürste; Schablonenpinsel

2264 step turner
[hyd/met]
f mandrin de pliage de plomb laminé
e mandril plegador de plomo laminado
i mandrino piegatore di piombo laminato
n bladloodbuigdoorn
d Bleiblechbiegedorn

2265 stippling
[dec]
f pointillage
e picado
i ritoccamento
n tamponneren
d Tupfen

2266 stippling brush
[dec]
f broche de pointillage
e brocha para picar
i pennello per ritoccare
n tamponneerkwast
d Tupfpinsel

* **stippling hammer → 1810**

2267 stippling iron
[art]
f fer à pointiller
e puntero
i subbia
n punteerijzer
d Punktiereisen

2268 stippling wheel
[art]
f roue à pointer
e rueda para puntear
i marcapunti
n stippelrad
d Punktierrad

2269 stock brush
[dec/hou]
f brosse mouilleuse
e brocha humectadora
i aspersorio
n vochtkwast
d Befeuchtungsbürste

* **stock shears → 214**

* **stone bolt → 766**

* **stone chisel → 1576**

* **stone dressing hammer → 471**

* **stonemason's saw → 1575**

2270 stone saw
[saw/sto]
f scie à pierre; scie de carrière
e sierra de cantero; sierra para piedra
i sega di cava di pietra
n steengroevezaag; steenzaag
d Steinmetzsäge; Steinsäge

2271 stone tongs
[pli]
f pince de levage
e tenaza elevadora
i tenaglie elevatori
n heftang
d Hebezange; Steinträgerzange

* **stone tongs → 1653**

2272 stop
[car/pla/saf]
f arrêt
e limitador
i arresto
n stop
d Haltevorrichtung

* **stopped mortise → 254**

2273 straddle milling
[fre]
f fraisage parallèle
e fresado con fresas acopladas; fresado paralelo
i fresatura parallela
n parallel frezen
d Parallelfresen

2274 straight-back blade
[saw]
f lame à dos rectiligne
e hoja con costilla rectilínea

i lama di sega con dorso rettilineo
n zaagblad met rechte rug
d Sägeblatt mit geradem Rücken

2275 **straight-cup grinding wheel**
[gri/san]
f meule boisseau cylindrique
e muela de copa cilíndrica
i mola coppa retta
n rechte komslijpschijf
d gerade Topfschleifscheibe

* **straight cylindrical awl → 1977**

* **straight edge → 2260**

2276 **straight-fluted reamer**
[dri/fre/rea]
f alésoir à cannelure droite
e escariador de estrías rectas
i alesatore a scanalatura diritta
n ruimer met rechte geslepen groeven
d geradegenutete Reibahle

2277 **straight-flute drill**
[dri]
f foret à rainures droites
e broca de estrías rectas; brocas de labios rectos
i punta di trapano a scanalatura diritta
n boor met rechte kanalen
d Bohrer mit geraden Nuten

2278 **straight gouge**
[chi]
f gouge droite
e gubia recta
i sgorbia retta
n rechte guts
d gerader Hohlbeitel

2279 **straight grinders**
[gri]
f meuleuses droites
e esmeriladoras rectas
i smerigliatrici rette
n horizontale slijpmachines
d Geradschleifer

* **straight grinding wheel → 1039**

2280 **straight halved joint**
[car/joi/woo]
f assemblage à mi-bois à trait droit
e empalme recto a media madera
i calettatura a mezzo legno; giunto a mezzo e mezzo
n rechte liplas
d gerades Blatt

2281 **straight lateral stake**
[cpt/for]
f fer à rabattre (droit); tranche droite de ferblanterie
e bigornia lateral recta
i bicornia laterale diritta
n rechte kantstaak
d Umschlageisen mit gerader Bahn

* **straight lock → 1939**

2282 **straight-peen-hammer**
[ham]
f marteau à panne droite
e martillo de peña recta
i martello a penna diritta
n rechte-penhamer
d Hammer mit gerader Pinne

2283 **straight shank drilling machine**
[dri]
f perceuse à tige cylindrique
e taladro de espiga cílindrica
i trapano a gambo diritto
n boormachine met cilindrische as
d Zylinderschaft-Bohrmaschine

2284 **straight tongue**
[car/joi/woo]
f languette droite
e lengüeta directa
i linguetta diretta
n rechte veer
d gerade Feder

* **straight toothing → 1051**

2285 **strainer**
[gen/sep]
f passoire
e zarranda
i colino
n zeefje
d Seiher; Sieb

* **strap → 303**

2286 **strap hinge**
[iwa]
f charnière à penture simple
e bisagra de ramal
i cerniera con occhiello
n plaatduimheng
d Fischband; Türangel

* **strap hinge → 1139**

2287 **strap wrench**
[spa]

f serre-tube à sangle
e llave de correa
i chiave a cinghia
n bandsleutel
d Bandschlüssel

* **straw roof hook → 1918**

* **streamline burner → 937**

2288 **stretcher rail**
[cpt/saw]
f lardon de serrage
e listón tensor
i staggio
n spanregel
d Spannleiste

2289 **stretching hammer**
[ham]
f marteau à dégrossir
e martillo para alargar
i martello tenditore
n rekhamer
d Streckhammer

2290 **strickle**
[agr/sha]
f chaploir
e afiladera del hoz
i affilatoio della falce messoria
n strekel
d Schnitter

2291 **strickle**
[fou]
f racloir
e terraja
i raschiatoio
n afstrijkmes; strijkmes
d Streichblech

2292 **strickle**
[mea]
f raclette
e rasero
i rasiera
n streek; strekel; strijkstok
d Streichholz

2293 **strickle board; striking board; sweep**
[fou]
f planche à trousser
e plantilla para aterrajar
i asse per sagomare
n trekplank
d Streichschablone

2294 **striding level**
[mea/sur]
f niveau principal d'un théodolite
e nivel a caballo
i livella principale di un teodolito
n buisniveau
d Röhrenlibelle

* **striking board → 2293**

2295 **striking button of a plane**
[cpt/pla]
f bouton-protecteur de rabot
e botón-protector de cepillo
i bottone di protezione di pialla
n slagknop
d Hobelknopf; Schlagknopf

* **striking tool → 335**

* **striking wedges → 839**

* **stripping knife → 1695**

2296 **stub drill**
[dri]
f foret à tige cylindrique
e broca de caña cilíndrica
i punta elicoidale con codolo cilindrico
n boor met cilindrische kolf
d Bohrer mit zylindrischem Schaft

* **stub tenon → 1463**

2297 **stud**
[iwa]
f clou doré
e clavo de decoración; clavo de roseta
i chiodo coronato
n sierspijker
d Ziernagel

2298 **stud**
[iwa]
f prisonnier
e espárrago
i prigioniero
n stokschroef
d Stiftschraube

2299 **stud**
[iwa]
f goujon; tourillon
e espiga
i traversino
n stift
d Stift

2300 **stud bolt**
[iwa]
f boulon; tire-fond
e viratillo
i bullone prigioniero
n schroefbout
d Schraubenbolzen; Stehbolzen

* **studio easel → 838**

2301 **stuffing box**
[eng/saf]
f presse-étoupe
e prensaestopas
i premistoppa
n pakkingbus
d Stopfbüchse

2302 **stump tenon**
[car/joi/woo]
f tenon conique
e muñón cónico
i tenone corto conico
n konische pen
d konischer Zapfen

* **subpress → 737**

* **subtense bar → 2246**

* **suction cleaner → 861**

* **surface cutter → 704**

* **surface ga(u)ge → 2070**

2303 **surface grinding; surfacing**
[gri]
f meulage à plat; surfaçage
e rectificado en plano
i molatura a piano
n vlakslijpen
d Flächenschleifen

2304 **surface-grinding machine**
[pol]
f rectificatrice de surfaces planes; surfaceuse
e rectificadora de superficies planas exteriores
i rettificatrice per superfici piane
n vlakschuurmachine
d Flächenschleifmaschine

* **surfacing → 2303**

* **surveyor's chain → 899**

2305 **surveyor's measuring tapes**
[mea/sur]
f mesures chantiers et arpentage
e cintas métricas para agrimensores
i rotelle metriche per agrimensori
n landmetermaten
d Geometerbandmasse; Landmesserbänder

2306 **surveyor's tape; wind-up tape measure**
[mea]
f mesure d'arpentage; mètre à ruban d'arpenteur
e metro de cinta de agrimensor
i metro di nastro di geometra
n landmeter
d Stahlbandmass mit umklappbarer Kurbel

2307 **swage block**
[gas]
f dé d'orfèvre; tas-étampe
e dado de orfebre
i dado per curvare; dado d'orafo
n ank
d Anke

* **swan neck mortise chisel → 1537**

2308 **swarf guide**
[cut/gen]
f couteau diviseur
e cuña de partir; guía de virutas
i guida di trucioli
n spouwmes
d Spaltkeil

2309 **Swedish pipe tongs**
[pli]
f pince à serre-tubes suédoise
e llave para tubos modelo sueco
i giratubi modello svedese
n Zweedse gaspijptang
d Schwedenzange

* **sweep → 2293**

2310 **sweep saw; turning saw**
[saw]
f scie à tourner
e sierra de cortar en vueltas; sierra de rodear
i sega a volgere
n draaizaag
d Rundsäge

2311 **swingle; swipple**
[agr]
f battoir
e mazo
i vetta

n vlegelknuppel
d Klöppel

* **Swiss pattern file → 1452**

* **synthetic fertilizer broadcaster → 967**

2312 **synthetic silicon carbide paper; wet and dry paper**
[gri/san]
f papier à carborundum
e papel abrasivo de carburo de silicio
i carta abrasiva a carburo di silicio
n carborundumpapier; waterproofschuurpapier
d Karborundpapier; wasserfestes Schleifpapier

T

2313 **T; tee**
[hyd/joi]
f raccord à T
e tubo en T
i ti
n T-stuk
d T-Stück

2314 **table**
[gen/tool]
f table de montage
e mesa de montaje
i tavola portapezzi
n montagetafel
d Montagetisch

2315 **tabled joint**
[joi/woo]
f assemblage à denture droite; emboîtement
e empalme endentado con doble contrejunta y pernos pesantes
i unione a dentatura retta
n lipverbinding
d ineinandergreifende Verzahnung

2316 **tacheometer; tachymeter**
[mea/sur]
f tachéomètre
e taquímetro
i tacheometro
n tachymeter
d Tachymeter

2317 **tachometer**
[eng/mea]
f tachymètre
e tacómetro
i tachimetro
n tachometer
d Drehzahlmessgerät; Tachometer

* **tachymeter → 2316**

2318 **tack**
[fas/iwa]
f semence
e tachuela
i semenza
n kopspijkertje
d Zwecke

2319 **tack hammer; upholsterer's hammer**
[ham/hou]
f marteau de matelassier; marteau de tapissier
e martillo de colchonero; martillo de tapicero
i martello di matassatore; martello di tappezziere
n matrassenmakershamer; stoffeerdershamer
d Polsterhammer; Tapezierhammer; Zweckenhammer

2320 **tackle**
[tra]
f palan
e garrucha; polea
i paranco
n takel
d Flaschenzug

2321 **tackle fall; tackle rope**
[cpt/tra]
f cordage de palan; corde de palan
e ramal de polea
i fune di paranco
n takeltouw
d Flaschenzugseil

2322 **tackle hook**
[cpt/tra]
f croc de palan
e gancho de polea
i gancio di paranco
n takelhaak
d Flaschenzughaken

* **tackle rope → 2321**

* **tack puller → 528**

* **taffrail log → 1715**

2323 **taft joint**
[hyd/joi]
f tube de plomb à bride
e tubo de plomo con brida
i tubo di piombo a flangia
n loodpijp met flens
d mit einem Flansch versehene Bleiröhre

2324 **tampin; turning pin**
[hyd]
f évasoir
e abocador
i allargatoio
n loodprop
d Bleirohraufweitdorn

2325 **tanged file**
[far]
f lime à soie
e lima con espiga

i lima a codolo
n vijl met adelaar; vijl met angel; vijl met arend
d Angelfeile

2326 tang of a file
[far]
f soie de lime
e espiga de lima
i codolo di lima
n adelaar van een vijl; angel van een vijl; arend van een vijl
d Feilenangel

2327 tap
[gen/hyd]
f robinet
e grifo
i rubinetto
n kraan
d Hahn

2328 tap
[scr]
f taraud à fileter
e macho de roscar
i maschio sbozzatore per filettare
n tap
d Gewindebohrer; Schraubenbohrer

2329 tape cramp
[car/joi/woo]
f presse à coller pour rubans
e sargento para cintas
i sergente per nastri
n bandlijmknecht
d Streifenleimzwinge

2330 taper cup grinding wheel
[gri/san]
f meule boisseau conique
e muela de copa cónica
i mola coppa conica
n konische komslijpschijf
d kegelige Topfschleifscheibe

2331 tapered roller bearing
[eng/tra]
f palier à rouleaux coniques
e cojinete de rodillos cónicos
i cuscinetto a rulli conici
n kegelrollenlager
d Kegelrollenlager

* **tapered slipstone → 1981**

2332 taper ground
[gri/sha]
f à affûtage concave
e de afilado cóncavo
i ad affilatura concava
n holgeslepen
d hohlgeschliffen

2333 taper pin
[iwa]
f goupille
e pasador cónico
i perno conico; spina conica
n konische pen
d Kegelstift; konischer Splint

* **taper pipe → 1421**

2334 taper shank drill
[cpt/dri]
f foret à queue conique
e broca con espiga cónica
i punta elicoidale con codolo conico
n boor met konische kolf
d Bohrer mit konischem Schaft

2335 taper tap
[scr]
f taraud ébaucheur; traçoir
e macho roscador cónico; primer macho de roscar
i maschio conico per filettare
n konische tap; voorsnijder
d Vorschneider

* **taper tap → 905**

2336 tap-out bar ***US*****; tapping bar** ***GB***
[fou/met]
f barre à piquer
e piquera
i barra per spillare
n afsteekijzer; afsteker
d Stichlochspiess

2337 tapping
[scr]
f taraudage
e roscado interior de agujeros
i filettatura interiore; maschiatura
n inwendig draadsnijden
d Innengewindebohren; Innengewinde-schneiden

* **tapping bar** ***GB*** **→ 2336**

2338 tap wrench
[spa]
f tourne-à-gauche
e llave giramachos; terraja de cojinete
i giramaschio

n wringijzer; wringstaaf
d Windeisen

2339 **tar brush**
[dec/saf]
f brosse à goudronner; guipon
e pincel para alquitranar
i pennello per incatramare; spalmatoio
n teerkop
d Teerpinsel; Teerquast

2340 **target rod**
[mea/sur]
f jalon à dispositif de lecture; voyant
e mira de corredera; mira de tablillo
i patina di traguardo
n meetlat met afleesschuif
d Messlatte mit Ableseschieber

2341 **T-bolt; tee bolt**
[iwa]
f boulon à tête en T
e perno con cabeza de muletilla
i bullone con testa a T
n T-bout
d Hammerkopfbolzen

* **tee → 2313**

2342 **tee hinge; T-hinge**
[iwa]
f penture T
e bisagra en T
i cerniera con occhiello in T
n kruisheng; T-scharnier
d T-Türangel

2343 **tee rest; T rest**
[lat/woo]
f poutrelle à T
e portador en T
i supporto a T
n T-steun
d T-Träger

2344 **tee slot cutter; T-slot cutter**
[fre]
f fraise à rainer en T
e fresa para ranurar en T
i fresa per scanalare in T
n T-groeffrees
d T-Nutenfräser

2345 **teeth doctoring; teeth level(l)ing**
[cpt/saw]
f lissure des dents
e nivelado de los dientes
i livellamento dei denti
n strijken van een zaag
d Gleichrichtung der Zähne

2346 **teeth doctoring iron; teeth level(l)ing iron**
[cpt/saw]
f outil de lissure
e útil nivelador
i arnese per livellare
n strijkijzer
d Gleichrichter

2347 **teeth level(l)ing**
[saw]
f alignement de dents; redressage de dents
e alineación de dientes
i allineamento di denti
n strijken van de zaagtanden
d Richten der Sägezähne

* **teeth level(l)ing → 2345**

* **teeth level(l)ing iron → 2346**

* **teeth setting iron → 2037**

2348 **teeth spacing**
[cpt/saw]
f gencive
e paso entre dientes
i spazio tra denti
n spatie tussen de tanden; tandtussenruimte
d Zahnzwischenraum

* **telephone hammer → 1746**

* **telephone pliers → 1533**

2349 **template; templet**
[mea]
f gabarit
e calibre; patrón; modelo
i sagoma
n mal; sjabloon
d Lehre; Schablone

* **template → 1234**

* **templet → 1234**

* **tenon and mortise joint → 1628**

2350 **tenon countersink**
[dri/rea]
f alésoir-fraise à chevilles
e avellanador de muñones
i accecatoio di tenoni

n penverzinkfrees
d Zapfensenker

2351 **tenon fence**
[cpt/dri/joi/wool]
f guide de tenonnage
e guía de muñón
i guida di tenone
n pengeleider
d Anschlag zum Zapfenschneiden

* **tenon saw → 123**

2352 **tensioning pliers**
[pli]
f pince tendeuse
e alicate(s) tensor(es)
i pinza tenditrice
n spantang
d Spannzange

* **thatched roof hook → 1918**

2353 **thatcher's hammer**
[ham]
f marteau de couvreur en chaume
e martillo de tejador
i martello di copritore di tetti di canne
n dekhamer
d Dachdeckerhammer

2354 **thatcher's hook**
[bui/cov]
f crochet de couvreur en chaume
e gancho de tejador de caña
i gancio di copritore di tetti di canne
n rietdekkershaak
d Rohrdeckershaken

2355 **thatcher's lath**
[bui/cov]
f latte de couvreur en chaume
e lata de tejador
i listello da copritore
n deklat
d Rohrdeckerslatte

2356 **thatcher's pushing board**
[bui/cov]
f planchette de couvreur en chaume
e tabla de tejador de caña
i tavola da copritore di canne
n rietdekkersdekspaan; drijfbord
d Rohrdeckersrichtbrett

2357 **thatching**
[bui/cov]
f couverture de chaume
e cobertura de caña; tejar con caña
i copertura di paglia
n rietdekken
d Rohrdeckung

2358 **thick pencil**
[mas]
f crayon de maçon
e lápiz de albañil
i matita da muratore
n metselaarspotlood
d Maurerbleistift

* **thimble → 2164**

* **T-hinge → 2342**

* **third tap → 297**

2359 **thistle lifter**
[agr/hor]
f échardonnoir
e extractor de cardos
i estirpatore di cardi
n distelhaak; distelschaar; distelsteker; disteltang; disteltrekker
d Distelstecher

2360 **thorn-hedge scissors**
[hor]
f cisaille de haie d'épines
e tijeras de seto de espinas
i forbici per siepi a spine
n doornhegschaar
d Dornheckenschere

* **thread-cutting chisel → 2054**

* **thread-cutting lathe → 2055**

2361 **thread grinding**
[scr]
f rectification de filets
e rectificador de roscas
i rettifica di filetti
n slijpen van schroefdraad
d Schleifen von Gewinde

* **threading die → 736**

* **threading file → 2068**

2362 **threading stock and die**
[scr]
f filière à fileter
e terraja de cojinetes de roscar
i filiera e maschio a filettare; filiera e matrice a filettare
n snijplaat met snijplaathouder
d Gewindeschneidkluppe

2363 thread rolling
[scr]
f laminage de filets
e laminación de roscas
i rullatura dei filetti
n walsen van schroefdraad
d Gewindewalzen

2364 three-bit drilling machine
[dri]
f perceuse à trois broches
e taladro a tres husillos; taladradora de tres portabrocas
i trapano a tre mandrini
n boormachine met drie spillen
d Dreispindelbohrmaschine

2365 three-flute cutter
[fre]
f fraise à trois rainures
e fresa de tres ranuras
i fresa a tre scanalature
n driezijdige frees
d Dreischneiderfräser

2366 three-jaw chuck
[cpt/dri]
f mandrin à trois mors
e mandril de tres garras
i mandrino a tre griffe
n boorhouder met drie bekken
d Dreibacken-Bohrfutter

2367 three-legged frame; three-legged stand; tripod
[cpt/gen]
f trépied
e tripode
i treppiede
n driepoot; statief
d Dreifuss; Stativ

2368 three-prong cultivator
[agr]
f griffe sarcleuse
e cultivador de tres dientes
i sarchiatrice a tre denti
n driebladshak
d Dreizinkgrubber; Jätekralle; Wühleisen

2369 three-square file; triangular file
[far]
f lime triangulaire; tiers-point
e lima triangular
i lima triangolare
n driekante vijl
d Dreikantfeile

2370 three-tined hoe
[agr/hor]
f fourche à fumier à trois dents
e horca de tres puntas para el estiércol
i forcone a tre denti per letame
n drietandsmestvork
d dreizinkige Mistgabel

2371 threshing floortools
[agr]
f outils de battage; outils de l'aire
e herramientas de la era
i arnesi dell'aia
n deelgereedschap; dorsgereedschap
d Dreschgeräte; Tennengeräte

* **throat of a plane → 906**

2372 through-cutting scissors
[she]
f ciseau passe-tôle
e tijeras de planchas
i cesoia da passo-lastra
n doorloopblikschaar; doorloopschaar
d Durchlaufschere

2373 through tenon
[car/joi/woo]
f tenon passant
e muñón pasante
i tenone passante
n doorlopende pen
d durchlaufender Zapfen

2374 thumbscrew
[fas/iwa/scr]
f papillon; vis à ailettes; vis ailée; vis à oreilles
e tornillo de mariposa; tornillo de orejetas
i vite con testa ad alette
n vleugelmoer
d Flügelschraube

2375 tide ga(u)ge
[mea/nav]
f marégraphe; marémètre; maréomètre
e mareógrafo; mareómetro
i mareografo; mareometro; scala di livelli
n peilschaal
d Peilstange

2376 tightening nut
[cpt/fas/iwa]
f écrou de blocage; écrou de serrage
e tuerca apretadora
i dado di serraggio
n spanmoer
d Klemmutter; Spannmutter

2377 tightening spanner
[cpt/spa]
f clé de blocage; clé de serrage
e llave apretador
i chiave di serraggio
n spansleutel
d Mutternschlüssel

2378 tile cutter; tile cutting machine
[dec/pli]
f coupe-carreaux; coupe-dalles
e corteazulejos; cortebaldosas
i tagliapiastrelle
n tegelsnijder
d Fliesenschneider

2379 tile pliers
[pli]
f pince à carreaux; pince à dalles
e alicate(s) para azulejos; alicate(s) para baldosas
i pinza fora piastrelle
n tegelkniptang
d Fliesenlochzange

* **tiling hammer → 332**

* **timber bob → 1542**

* **timber janker → 1542**

2380 tin file
[far/met]
f lime à étain
e lima de estaño
i lima a stagno
n tinvijl
d Zinnfeile

2381 tinman's hammer; tinsmith's hammer; whitesmith's hammer
[ham/met]
f marteau d'affineur; marteau de ferblantier; marteau de tôlier
e martillo de hojalatero
i martello di lattoniere
n blikslagershamer
d Klempnerhammer

* **tinman's shears** *US* **→ 1784**

2382 tip
[cut/gen/tool]
f pointe
e punta
i punta
n punt
d Spitze

2383 toe of a plane
[cpt/pla]
f corne de rabot
e punta de cepillo
i nasello di pialla
n neus van een schaaf
d Nase eines Hobels

2384 toggle
[cpt/saw]
f tendeur
e husillo tensor; palanca tendora
i nottola
n spanner
d Spanner; Spannspindel

2385 tolerance
[gen/mea]
f tolérance
e tolerancia
i tolleranza
n tolerantie
d Toleranz

2386 tommy bar
[cpt/spa]
f broche à visser
e barra de maniobra; espiga para llave de tubo; palanca de mano
i asta per chiave a tubo
n draaistift
d Drehstift

* **tongs → 1652, 1744**

2387 tongue
[chi/cpt/tool]
f époinçon; soie
e cola; espiga
i ghiera
n adelaar; angel; arend; staart
d Angel

* **tongue and groove joint → 1215, 1786**

2388 tongue plane
[car/pla]
f bouvet mâle
e cepillo de machos
i pialla a linguette
n messingploeg; messingschaaf
d Federhobel

2389 tongue-shaped trowel
[mas]
f truelle en forme de langue
e paleta lengüeforme
i cazzuola in forma di maschio

n pannenstrijker
d Zungenkelle

* **tonguing plane → 1785**

2390 tool angle
[cpt/gen/tool]
f angle de taillant
e ángulo de corte
i angolo di taglio
n gereedschapshoek
d Werkzeugwinkel

2391 tool approach angle; tool lead angle
[gen/tool]
f angle de direction complémentaire de l'outil
e ángulo de dirección complementario del útil
i angolo di direzione complementario dell'utensile
n complementaire instelhoek
d Einstell-Ergänzungswinkel

2392 tool assembly
[gen/tool]
f ensemble outil
e conjunto útil
i complesso utensile
n gereedschapsopbouw
d Werkzeugaufbau

2393 tool bag
[tool]
f sac à outils; trousse d'outils
e bolsa de herramientas
i sacco d'utensileria
n gereedschapstas
d Werkzeugtasche

2394 tool board
[gen/tool]
f panneau à outils
e tablero de herramientas
i tavola d'utensileria
n gereedschapsbord
d Werkzeugbrett

2395 tool box; tool case
[gen/tool]
f boîte à outils; caisse à outils; nécessaire à outils
e caja de herramientas
i cassetta d'utensileria
n gereedschapskist
d Werkzeugbüchse

2396 tool car
[gen/tool]
f chariot à outils; chariot porte-outil
e carretilla de herramientas
i carro d'utensileria
n gereedschapswagen
d Werkzeugwagen

2397 tool carrier slide
[gen/lat/tool]
f chariot porte-outil
e carro de portaútil
i carrello di portautensile
n beitelhouderslede
d Meisselhalterschlitten

* **tool case → 2395**

2398 tool chest
[gen/tool]
f armoire à outils
e armario de herramientas
i armadio per utensileria
n gereedschapskast
d Werkzeugschrank

2399 tool clamp
[fas/gen/tool]
f agrafe porte-outil
e sujetador de herramientas
i morsa utensileria
n gereedschapsklem
d Werkzeughalteklamme

* **tool clapper → 527**

2400 tool cutting edge angle
[cut/gen/tool]
f angle de direction d'arête outil
e ángulo de dirección de filo del útil
i angolo di direzione del tagliente
n snijkantshoek
d Einstellwinkel

2401 tool cutting edge inclination
[cut/gen/tool]
f angle d'inclinaison d'arête d'outil
e ángulo de inclinación del canto cortante
i angolo d'inclinazione del lato tagliente
n snijkantshoekhelling
d Neigungswinkel

2402 tool cutting edge plane
[cut/gen/tool]
f plan d'arête d'outil
e plano del canto cortante
i piano del lato tagliente
n snijvlak van het gereedschap
d Werkzeugschneideebene

2403 **tooler**
[ham/sto]
f boucharde
e bujarda
i martello da tagliapietre
n boucharde; bouchardeerhamer
d Steinhauerkrönel

2404 **tool failure**
[gen/tool]
f mort d'outil
e defecto de útil
i difetto d'utensile
n uitval van gereedschap
d Erliegen

2405 **tool grinder**
[gen/sha/tool]
f affûteuse d'outils
e afiladora para útiles
i affilatrice per utensili
n gereedschapsslijpinrichting
d Schärfvorrichtung für Werkzeuge; Schleifvorrichtung für Werkzeuge

2406 **tool grinding**
[gen/sha/tool]
f affûtage d'outils
e afilado de útiles
i affilatura d'utensili
n slijpen van het gereedschap
d Werkzeugschleifen

2407 **tool holder** *GB*; **tool post** *US*
[cpt/tool]
f porte-outil
e portaútil
i portautensile
n gereedschapshouder
d Werkzeughalter

* **tool lead angle → 2391**

2408 **tool life**
[gen/tool]
f durée de l'outil
e vida del útil
i vita dell'utensile
n levensduur van het gereedschap
d Standvermögen; Standzeit

2409 **toolmaker**
[gen]
f outilleur
e constructor de herramientas; mecánico herramentista
i attrezzista
n gereedschapsmaker
d Werkzeugmacher

2410 **tool offset**
[gen/tool]
f décalage d'outil
e desviación de útil
i scostamento dell'utensile
n gereedschapsverschuiving
d Werkzeugversatz

2411 **tool outfit; tool set**
[gen/tool]
f jeu d'outils
e juego de herramientas
i utensileria
n stel gereedschappen
d Werkzeugsatz

2412 **tool post**
[cpt/lat/tool]
f étrier d'un tour
e portacuchilla; portaútil cortante
i portacoltello; portautensile da taglio
n snijijzerhouder
d Schneidschabhalter

* **tool post** *US* **→ 2407**

* **tool set → 2411**

2413 **tool shop**
[gen/tool]
f atelier d'outillage
e taller de construcción de útiles
i officina di costruzione d'utensili
n gereedschapsmakerij
d Werkzeugmacherei

2414 **tool steel**
[gen]
f acier à outils
e acero de herramientas
i acciaio per utensili
n gereedschapsstaal
d Werkzeugstahl

* **tooth chisel → 529**

2415 **toothed chisel**
[chi]
f ciseau à tranchant denté
e cincel de filo dentado
i scalpello a filo dentato
n beitel met getande snede; kambeitel
d Beitel mit gezahnter Schneide; Zahneisenmeissel

2416 **toothed half-lap joint; toothed halving**
[car/joi/woo]
f assemblage denté à mi-bois
e ensambladura dentada a media madera

i giunto dentato a mezzo legno
n halfhoutverbinding met tanden
d gezahnte Halbholzverbindung

2417 **toothed tongs**
[for/pli]
f tenaille(s) à gueule de loup
e tenazas de boca de lobo
i tenaglia a gola de lupo
n smeedtang met wolfsmuilbekken
d Wolfsmaulzange

2418 **toothed wheel**
[eng/gen/tra]
f roue dentée
e piñón; rueda dentada
i ruota dentata
n tandrad
d Zahnrad

* **toothed wheel puller → 552**

2419 **toothing plane**
[pla]
f rabot fer à dents
e cepillo con hierro dentado
i pialla a ferro dentato
n tandschaaf
d Zahnhobel

* **top cutting nippers → 894**

* **top cutting pliers → 584**

2420 **top die**
[for]
f matrice supérieure
e matriz superior
i matrice superiore
n bovenmatrijs
d oberer Schmiedesattel

2421 **top swage**
[cpt/for]
f étampe dessus à queue
e estampa superior
i stampo superiore
n bovenstempel
d Obergesenk; Rundgesenk oben

2422 **torque spanner; torque wrench**
[spa]
f clé dynamométrique
e llave con par de apriete prefijado; llave con dinamómetro de torsión
i chiave dinamometrica
n draaimomentsleutel; momentsleutel; torsiesleutel
d Drehmomentschlüssel

* **tower pincers → 1173**

2423 **towing line**
[saf/tra]
f câble de remorque; guide-rope
e cable de remorque
i cavo di rimorchio
n sleepkabel; sleeplijn
d Schlepptau

2424 **traditional sharpening**
[saw]
f affûtage traditionnel
e afilado tradicional
i affilatura tradizionale
n klassiek slijpen
d normales Schleifen

* **trammel → 184**

2425 **transparent protective shield**
[gen/gri/pla/saf]
f écran transparent de protection
e rejilla transparente protectora
i piastra trasparente di protezione
n slijpscherm; veiligheidsscherm
d Augenschutz aus Sicherheitsglas

2426 **traverse drilling machine**
[dri]
f perceuse de traverses
e taladradora ranuradora; taladro de ajuste lateral
i trapano di traverse
n dwarsboormachine
d Querbohrer; Schwellenbohrer

* **treadle blower → 1082**

* **tree brush → 157**

2427 **tree cleaning brush**
[cle/foy]
f brosse métallique d'arbres; décrotteuse d'arbres
e cepillo de alambre para trozas
i spazzola metallica per alberi
n boomborstel
d Baumdrahtbürste; Baumreinigungsbürste

2428 **tree ladder**
[foy]
f échelle d'arbres; échelle double
e escalera de árboles
i scala d'alberi
n boomladder
d Baumleiter

* **tree pruner → 109**

* **tree scraper → 162**

2429 trenail; trunnel
[car/joi/wool]
f cheville de bois; fenton
e cabilla; clavija de madera; cuña de llave
i caviglia di legno
n houten pen
d Holznagel

* **T rest → 2343**

2430 triangular bladed draw hoe
[agr/hor]
f binette à lame triangulaire; houe à lame triangulaire
e almocafre con pala triangular; azadón con pala triangular
i sarchio a lama triangolare
n driehoekige hak
d Dreieckziehhacke

* **triangular file → 2369**

2431 triangular sharpening stone
[gri/sha]
f pierre à affûter triangulaire
e piedra de amolar triangular
i pietra da affilare triangolare
n driekantaanzetsteen
d Dreikantschleiffeile

* **triangular trowel → 59**

* **triblet → 1944**

2432 trident spear
[nut]
f foène; foine; trident de pêcheur
e arpón para pescar; tridente de pesca
i tridente di pescatore
n drietand
d dreizackiger Fischspiess

2433 trimmer
[cle/met]
f ébarbeur
e desbarbadora
i sbavatore
n afbaarder
d Entgrater

2434 trimmer joint
[car/joi/wool]
f assemblage à tenon renforcé
e ensambladura con muñón de refuerzo
i incastro con tenone rinforzato
n versterkte penverbinding
d Verbindung mit versetztem Zapfen

2435 trimming axe
[axe/foy]
f épaule de mouton; hache d'équarrissage
e hacha acabadora
i ascia di squadratura
n beslagbijl; slechtbijl
d Zurichtaxt

2436 trimming grinding wheel
[cle/cpt/gri]
f disque d'ébarbage; disque abrasif pour nettoyer
e muela desrebarbadora; muela rebarbadora
i disco rifinitore
n afbraamschijf; bijwerkschijf
d Abgratscheibe; Schruppscheibe

* **trimming knife → 2127**

* **trimming saw → 2156**

* **triple last → 1365**

* **tripod → 2367**

2437 trommel
[eng/gen/min]
f trieur; trommel
e criba giratoria
i vaglio rotante a tamburo
n trommelzeef
d Trommelsieb

2438 trowel
[mas]
f plâtroir; taloche; truelle de maçon
e llana de albañil; paleta de albañil; palustre
i cazzuola di muratore
n metseltroffel; troffel
d Maurerkelle

2439 trowel with flexible blade
[mas]
f truelle à lame flexible
e paleta de hoja flexible
i cazzuola a lamina flessibile
n bandtroffel; troffel met buigzaam blad
d Kelle mit Biegeblatt

2440 true size reamer
[dri/fre/rea]
f alésoir de précision
e escariador de precisión
i alesatore di precisione
n precisieruimer
d Präzisionsreibahle

* **trunnel → 2429**

* **trying plane → 1466**

* **try plane → 1466**

2441 **try square**
[car/mea]
f équerre épaulée; équerre de menuisier
e escuadra de carpintero
i squadra a battente; squadra da falegname
n blokhaak; schrijfhaak; winkelhaak
d Anschlagwinkel; Tischlerwinkel; Zimmermannswinkel

* **try square → 124**

* **T-slot cutter → 2344**

2442 **tub**
[che/coo/gen]
f bac; baquet; cuve
e herrada; tino
i botte; mastello; tino
n kuip; tobbe
d Bütte; Kübel; Küppe; Wanne

* **tube cutting apparatus → 1753**

2443 **tubular rivet**
[iwa/riv]
f rivet tubulaire
e remache hueco; remache tubular
i ribattino tubolare
n holle klinknagel
d Hohlniet

2444 **tungsten carbide paper**
[gri/san]
f papier à métal dur
e papel abrasivo revestido de carburo de tungsteno
i carta abrasiva a carburo di tungsteno
n hardmetaalschuurpapier
d Hartmetallschleifpapier

2445 **tungsten-carbide tipped**
[gen/tool]
f mise rapportée en carbure de tungstène
e chapado con carburo de tungsteno; con puntas de carburo de tungsteno
i con punte al carburo di tungsteno
n voorzien van hardmetaallaag
d hartmetallbestückt

* **tup → 1898**

* **Turk's head → 444**

* **turnbuckle → 2065**

* **turner's chisel → 345**

2446 **turner's gouge**
[chi/lat/woo]
f gouge de tourneur
e gubia de tornero
i sgorbia di tornitore
n draaiguts
d Drechslerhohlbeitel; Drehröhre zum Vordrehen

2447 **turning**
[lat]
f tournage
e torneado
i tornitura
n draaien
d Drehen

* **turning pin → 2324**

2448 **turning radius**
[she]
f rayon de giration
e radio de giro; radio de viraje
i raggio di sterzo
n draaiingsstraal
d Kurvenradius

2449 **turnings**
[lat/met]
f copeaux de tournage
e torneadura; virutas de torno
i trucioli torniti
n draaispanen
d Drehspäne

* **turning saw → 2310**

* **turning tools → 1502**

2450 **turret**
[lat]
f tourelle revolver
e cabezal de revólver; portaherramienta revólver; torreta de revólver
i testa a revolver; torretta a revolver
n revolverkop
d Revolverkopf

2451 **turret lathe**
[lat]
f tour revolver
e torno revólver
i tornio revolver
n revolverdraaibank
d Revolverdrehmaschine

2452 tusk tenon
[car/cpt/joi/woo]
f tenon à renfort
e muñón de refuerzo
i tenone rinforzato
n versterkte pen
d versetzter Zapfen

* **twist bit → 98, 793**

2453 twist gimlet
[dri]
f vrille conique
e barrena cónica
i succhiello conico
n konische fretboor; konische krukboor
d konischer Nagelbohrer

2454 two-bit drilling machine
[dri]
f perceuse à double broche
e taladradora de dos portabrocas; taladro a dos husillos
i trapano a due mondrini
n boormachine met twee spillen
d Zweispindelbohrmaschine

* **two-ferrule-handle → 1290**

2455 two-handed cross-cut saw; woodman's saw
[foy/saw]
f scie à deux mains; scie passepartout
e sierra de dos manos; tronzador
i sega diritta a due mani
n boomzaag; trekzaag
d Zugsäge

2456 two-jaw chuck
[cpt/dri]
f mandrin à deux mors
e mandril de dos garras
i mandrino a due griffe
n boorhouder met twee bekken
d Zweibacken-Bohrfutter

* **two-men cross-cut saw → 756**

* **two-start thread → 760**

2457 tyre fork lever
[tra]
f démonte-pneu à fourchette
e desmontable de pie de cabra; palanca de pie de cabra
i leva smontapneumatici a forca
n gevorkte bandenlichter
d Kniefuss; Kuhfuss

2458 tyre lever
[tra]
f démonte-pneu
e desmontable de cubiertas; palanca de cubiertas
i leva smontapneumatici
n bandafnemer; bandlichter; bandwipper
d Reifenheber

2459 tyre spreader
[tra]
f écarte-pneu
e separador de cubiertas
i allargacerchi
n bandenspreider
d Reifenspreize

U

2460 **U-bolts**
[iwa/tra]
f boulon forme U
e perno en U con rosca en ambos extremas
i staffa per fissaggio delle molle a balestra
n U-vormige bout
d U-förmiger Bolzen

2461 **U-clamp; U-pattern clamp**
[cla/iwa]
f bride de serrage en U
e brida de sujeción en U
i brida di serraggio in U; morsa ad U; staffa per fissaggio
n U-modelklem
d Gabelspanneisen

2462 **U-leather; U-packing**
[fas/saf]
f garnissage en U
e guarnición en U
i guarnizione ad U
n U-pakking
d U-Packung

2463 **ultrasonic machining**
[dri/eng]
f usinage ultrasonique
e maquinado por vibración ultrasónica
i lavorazione a macchina con vibrazioni ultrasoniche
n ultrasone machinale behandeling
d maschinelle Ultraschall-Bearbeitung

2464 **underwater saw**
[saw]
f scie à recéper sous eau
e sierra para cortar bajo agua
i sega per tagliare sotto acqua
n grondzaag; zaag voor het koppen van palen
d Unterwassersäge

2465 **undulating toothing; wavy set**
[saw]
f denture ondulée
e dentadura ondulada
i dentatura ondulata
n golvende vertanding
d gewellte Zahnung

2466 **unified screw thread**
[scr]
f filetage normalisé
e roscado patrón
i filettatura campione
n genormaliseerde schroefdraad
d normalisiertes Gewinde

2467 **unilateral tolerance**
[eng]
f tolérance unilatérale
e tolerancia unilateral
i tolleranza unilaterale
n eenzijdige tolerantie
d einseitige Toleranz

2468 **union**
[fas/hyd]
f manchon de raccord; union
e racor; unión
i hocchettone; raccordo
n verbindingsbuis
d Kupplung; Rohrverbindung

* **universal chuck → 2071**

2469 **universal grinder**
[pol]
f meuleuse universelle
e esmeriladora universal
i smerigliatrice universale
n universele slijpmachine
d Allzweckschleifmaschine; Universal-schleifmaschine

* **universal knife → 2127**

2470 **universal milling machine**
[fre]
f fraiseuse universelle
e fresadora universal
i fresatrice universale
n universele freesmachine
d Universalfräsmaschine

2471 **universal plane**
[pla]
f rabot universel
e cepillo universal
i pialla per tutti gli usi
n universele schaaf
d Allzweckhobel

* **universal scissors → 44**

2472 **unscrewing**
[scr]
f dévissage
e destornillamiento
i svitatura
n uitschroeven
d Losschrauben

* **U-packing → 2462**

* **U-pattern clamp → 2461**

* **upholsterer's hammer → 2319**

2473 **upper case**
[pri]
f haut de casse
e caja alta
i cassa di sopra; maiuscolo
n bovenkast
d oberer Schriftkasten; Oberkasten

2474 **upper grip**
[cpt/dri]
f poignée supérieure
e mango superior
i manico superiore
n bovenhandgreep
d oberer Griff

V

* **vacuum cleaner → 861**

2475 **varnish brush**
[dec]
f brosse de vernissage
e brocha para barnizar
i pennello per vernice
n verniskwast
d Firnispinsel

2476 **varnished wooden handle**
[cpt/too]
f poignée en bois vernie
e empuñadura de madera barnizada
i manico in legno verniciato
n geverniste houten handgreep
d poliertes Holzheft

* **vat → 163**

2477 **Vee joint**
[car/joi/woo]
f assemblage en V de planches
e ensambladura en V de tablas
i incastro in V di asse
n puntschroot; V-schroot
d V-Verbindung von Brettern

2478 **Vee thread**
[iwa/scr]
f filetage en V; filetage triangulaire
e rosca en V; rosca triangular
i filettatura a V; filettatura triangolare
n driehoekige schroefdraad
d Dreieckschraubengewinde

2479 **veiner**
[chi]
f gouge à lame en U
e gubia de hoja en U
i sgorbia a lama in U
n guts met U-vormig blad; smalle heelholle guts
d U-Hohlbeitel

2480 **veining bit; veining cutter**
[car/fre/woo]
f fraise pour rainure demironde
e fresa para acanalar
i fresa per scanalare
n halfronde groeffrees
d Halbrundnutenfräser

2481 **veining brush**
[dec]
f brosse de veinage; peigne à faux bois
e pincel de veteado; pincel para vetear
i pennello da venatura
n aderkwast
d Maserpinsel

* **veining cutter → 2480**

2482 **veneer**
[car/joi/woo]
f placage
e chapa de madera
i impiallacciatura
n fineer
d Furnier

2483 **veneer cutting machine**
[car/cut/joi/woo]
f coupe-placage; machine à trancher le bois
e fraccionador de chapa de madera
i tagliatrice per impiallacciatura
n fineersnijmachine
d Furnierschneider

2484 **veneer facing; veneer sheet**
[car/joi/woo]
f feuillet de placage
e hoja de enchapado
i foglio da impiallacciatura
n fineerblad
d Furnierblatt

2485 **veneering**
[car/joi/woo]
f placage
e enchapado
i impiallacciatura
n fineren
d Furnierung

2486 **veneering hammer**
[ham/woo]
f marteau à placage
e martillo de enchapado; martillo para enchapar
i martello per impiallacciatura
n belijmhamer; belijmingshamer; fineerhamer; plakhamer
d Furnierhammer

2487 **veneering press**
[car/joi/woo]
f presse à plaquer
e prensa de enchapado
i pressa per compensato; pressa per impiallacciatura
n fineerpers
d Furnierpresse

2488 veneering rack
[car/joi/wool]
f châssis de placage
e caballete de enchapado
i intelaiatura per impiallacciatura
n fineerbok
d Furnierbock

2489 veneering wood
[car/wool]
f bois à plaquer; bois de placage
e madera de chapa
i legno da impiallacciatura
n fineerhout
d Furnierholz

2490 veneer plane
[car/pla/wool]
f vastringue de placage
e cepillo rasquetador para enchapadas; chapeadora
i pialla per impiallacciatura
n fineerschaaf
d Furnierschaber; Furnierschabhobel

2491 veneer planing machine
[car/joi/pla/wool]
f raboteuse de placage
e acepilladora para chcapa de madera
i piallatrice per impiallacciatura
n fineerschaafmachine
d Furnierhobelmaschine

2492 veneer saw
[saw]
f scie à placage
e sierra de enchapado
i sega di impiallacciatura
n fineerzaagje; stootbordenzaagje
d Furniersäge

2493 veneer sawing mill
[car/joi/saw/wool]
f scierie de bois de placage
e aserradero de madera de chapa
i segheria per legno d'impiallacciatura
n fineerhoutzagerij
d Furnierholzsägewerk

2494 veneer scissors
[car/cut/joi/wool]
f cisailles à placage
e tijeras para enchapado
i cesoie per impiallacciatura
n fineerschaar
d Furnierschere

*** veneer sheet → 2484**

2495 veneer side cutting machine
[car/cut/joi/wool]
f machine à cisaillements des bords de placage
e máquina para recortar los bordes de chapa
i macchina per tagliare i lati d'impiallacciatura
n fineerkantensnijmachine
d Furnierkanten-Beschneider

2496 ventilated backing pad
[cpt/gri]
f plateau souple ventilé
e plato flexible ventilado
i piattorello flessibile ventilato
n poreus elastisch schuurblad
d belüfteter elastischer Schleifteller

2497 vernier
[gen/mea]
f démultiplicateur; vernier
e nonio
i nonio; verniero
n nonius
d Feineinsteller; Nonius

2498 vernier arm
[mea/sur]
f bras du vernier
e brazo de nonio
i braccio del verniero
n noniusarm
d Noniusarm

2499 vernier cal(l)ipers
[mea]
f pied à coulisse vernier
e calibre de nonio; pie de rey con nonio
i calibro a nonio; calibro a verniero
n schuifmaat met nonius
d Schublehre mit Nonius

*** vernier cal(l)ipers → 2168**

2500 vernier depth ga(u)ge
[mea]
f calibre à coulisse de profondeur
e calibre para profundidad
i calibro per profondità
n diepteschuifmaat
d Tiefenlehre; Tiefenmass

2501 vertical drilling machine
[dri]
f perceuse verticale
e taladro vertical
i trapano verticale

n vertikale boormachine
d Vertikalbohrmaschine

* **vertical log frame** *GB* → **1125**

2502 vertical miller; vertical milling machine; vertical spindle milling machine
[fre]
f fraiseuse verticale
e fresadora vertical
i fresatrice verticale
n vertikale freesmachine
d Senkrechtfräsmaschine

2503 V-groove bit; V-groove cutter
[car/fre/woo]
f fraise 90°
e fresa de ranura en V
i fresa a scanalatura in V
n V-groeffrees
d V-Nutenfräser

* **vial** → **1522**

2504 vibrating sander attachment
[cpt/san]
f adaptation ponceuse vibrante
e aditamento para lijadora vibrante
i adattamento per levigatrice vibrante
n voorzetstuk voor vibrerende schuurmachine
d Schwingsandervorsatz; Schwingschleif-vorsatz

* **vice-grip wrench** → **1624**

2505 vitrified grinding disk in corundum
[cpt/gri]
f meule vitrifiée en corindon
e muela abrasiva vitrificada de corindón
i mola abrasiva vetrificata al corindone
n korundslijpschijf met keramische binding
d Korundschleifscheibe mit keramischer Bindung

2506 V jointer
[joi/mas]
f truelle en V à joints de briques
e paletín en V
i ferro in V per giunti di mattoni
n V-voegspijker
d V-förmige Fugenkelle

2507 vole trap
[agr/hor/saf]
f piège à mulots
e trampa para ratones
i trappola per arvicole
n woelmuisval
d Wühlmausfalle

* **voltage detector** → **1561**

* **voltage indicator** → **1561**

* **V tool** → **1713**

W

* **wad punch → 83**

* **wallboard saw → 576**

2508 **wall hook**
[iwa]
f crochet mural
e alcayata; escarpia
i gancio a muro
n muurhaak; puntduim
d Mauerhaken; Wandhaken

* **warding file → 1471**

2509 **washbasin nut pliers**
[hou/pli]
f pince de devant à écrous pour robinets de montée
e alicate(s) para tuercas de robinetes
i pinza per dadi rubinetti
n wastafelmoertang
d Standhahnmutterzange; Standhahnzange

2510 **washer**
[iwa]
f rondelle
e arandela
i ranella; rondella
n borgring; sluitring
d Unterlegscheibe

* **washer cutter → 1378**

2511 **watchmaker's chisel**
[chi/wat]
f ciseau d'horloger
e broca de relojero; cojinete de terrajar para relojero
i scalpello di orologiaio
n boorbeitel
d Uhrmachermeissel

2512 **watchmaker's screwdriver**
[scd/wat]
f tournevis d'horloger
e destornillador de relojero
i cacciavite di orologiaio
n horlogemakersschroevedraaier
d Uhrmacherschraubendreher

2513 **watchmaker's tweezers**
[pli/wat]
f brucelles d'horloger
e pinzas de relojero
i pinzette di orologiaio
n korentang
d Kornzange

* **watering can** *GB* **→ 2238**

2514 **water pump pliers**
[pli]
f pince multiprise réglable
e tenacilla poliasidoras
i pinza regolabile per dadi e tubi
n waterpomptang
d Wasserpumpenzange

2515 **water torch**
[cle/too]
f chalumeau à couper à jet d'eau à haute pression
e soplete para cortar mediante chorro de agua de alta presión
i cannello per tagliare mediante getto di acqua ad alta pressione
n hoge-druk-waterstraalsnijder
d Hochdruck-Wasserstrahlschneiddüse

* **wavy set → 2465**

2516 **wear of grinding wheel**
[cpt/gri]
f usure de meule
e desgaste de muela
i usura di mola
n slijtage van de slijpschijf
d Verschleiss der Schleifscheibe

2517 **weatherometer**
[dec/mea]
f appareil de mesure de la désintégration
e medidor de disgregación
i decolorimetro
n verweringsmeter
d Verwitterungsmesser

2518 **weaver's pliers**
[pli/tex]
f pince de tisseur
e alicate(s) de tejedor
i pinza di tessitore
n weverstang
d Weberzange

2519 **webbing hammer**
[ham/hou]
f marteau à sangles
e martillo de cinchas
i martello per cinghie
n singelhamer
d Gurthammer

2520 wedge
[cpt/gen]
f coin
e cuña
i cono; cuneo; zeppa
n wig
d Keil

* **wedge → 79**

2521 wedge of a plane
[cpt/pla]
f coin de rabot
e cuña de cepillo
i cuneo di pialla
n keghout; opsteker; spie van een schaaf
d Hobelkeil

2522 weeder
[agr]
f échardonnoir
e desherbador
i estirpatore
n onkruidsteker
d Unkrautstecher

2523 weeding hoe
[agr/hor]
f binette sarcleuse; croc à sarcler; sarcloir
e legón; ligón; sacho
i zappa per sarchiare
n schoffel
d Jätehäckchen

2524 weightometer
[mea/min]
f bascule enregistreuse
e báscula registradora
i basculla registatrice
n registrerende bascule
d registrierende Brückenwaage

* **welder's chipping hammer → 484**

2525 welding pliers
[pli/wel]
f pince à souder
e horquilla para soldar; mordazas para soldar
i pinza per saldare
n lasklemtang
d Schweissbordzange

* **welding rod → 988**

2526 welding transformer
[wel]
f transformateur de soudage
e transformador de soldadura
i trasformatore di saldatura
n lastrafo; lastransformator
d Schweisstransformator

2527 welt cutter
[lea]
f tranchet à faire les bords
e desvirador
i tagliaorli da cuoio
n kantsnijmes
d Randmesser

* **welt pincers → 1513**

* **wet and dry paper → 2312**

2528 wet drilling
[dri/min]
f perçage à eau
e perforación con injección de agua
i perforazione ad umido
n nat boren
d Nassbohren

2529 wet grinding
[gri]
f rectification à eau
e rectificación ad agua
i rettifica ad umido
n nat slijpen
d Nassschleifen

* **wheelbarrow → 166**

2530 wheel hub puller
[pul/tra]
f extracteur de moyeu de roue
e descalador de bujes
i estrattore di mozzi della ruota
n wielnaaftrekker
d Radnabenabzieher

2531 wheel puller
[pul/tra]
f extracteur de roue
e sacaruedas
i estrattore per ruote; levaruota
n wieltrekker
d Radabzieher

2532 wheel testing hammer
[ham/tra]
f marteau pour le contrôle des roues
e martillo comprobador de ruedas
i martello da verificatore
n treinsmidshamer
d Radprüfhammer

2533 **wheelwright's file**
[far/tra]
f lime de charron
e lima de carretero
i lima di carradore
n wagenmakersvijl
d Karosseriefeile

* **whetstone → 216, 1387**

* **whetstone sheath → 1388**

2534 **whetting anvil**
[agr]
f chaploir; enclumette à battre les faux
e yunque del guadañero
i incudine per affilare
n bootschaar; haarspit; spit
d Dengelamboss

2535 **whetting hammer**
[agr/ham]
f marteau à battre les faux
e martillo de afilar
i martello per affilare
n haarhamer
d Dengelhammer

* **whip saw → 1951**

* **whitesmith's hammer → 2381**

2536 **whitewash brush**
[dec]
f brosse à plafond; brosse à pocher
e brocha de encalar
i pennello per calce
n witkwast
d Kalkpinsel

* **whitewash brush → 743**

2537 **wimble**
[dri/min]
f tarière à glaise
e barrena de gusano
i trapano da argilla; trapano da loto
n leemboor
d Lehmbohrer; Schoppenbohrer

* **wimble → 97, 313, 1160**

2538 **winch**
[tra]
f treuil
e manivelo; torno
i verricello
n lier
d Winde

* **window brush → 2013**

* **wind-up tape measure → 2306**

2539 **wing compasses**
[mea]
f compas quart de cercle
e compás de espesores con arco
i compasso per spessori ad arco
n boogpasser; vleugelpasser
d Bogenzirkel

* **wing nut → 378**

2540 **winnow**
[agr]
f van
e arel; harnero
i vaglio
n wan
d Wanne

* **wire brush → 2257**

* **wire brush with scraper → 2049**

2541 **wire cutter**
[iwa/pli]
f coupe-fil; pince monseigneur
e cortaalambre
i pinza tagliafili
n draadkniptang
d Drahtschneider

2542 **wire-cutting scissors**
[pli/sci]
f pince à ciseaux; pince coupante
e alicate(s) de tijeras
i pinza a cesoie
n schaartang
d Scherenzange

2543 **wire nail**
[fas/iwa]
f pointe en fil à fer
e punta de hierro
i punta filo ferro
n draadnagel
d Drahtnagel

2544 **wire rake**
[agr/hor]
f râteau à feuilles
e escoba de alambres; escoba para la hojarasca
i rastrello a scopa; rastrello da fogliame
n bladhark
d Fächerbesen; Laubbesen

2545 wire reel
[ele/gen]
f dévidoir de fil
e devanadora de alambre
i mulinello di filo
n snoerhaspel
d Drahthaspel

* **wire rotary brushes → 1970**

* **wire saw → 392**

2546 wire strainer
[iwa]
f raidisseur; tendeur de fils
e templador; tensor de alambre
i tenditore di fili
n draadspanner
d Drahtspanner

2547 wire stripper; wire stripping pliers
[ele/pli]
f pince à dénuder
e tenazas para desmantelar hilos
i pinza spellafili
n draadstriptang; isolatie-afstriptang
d Abisolierzange

* **wire stripping pliers → 1437**

* **wobbling saw → 822**

2548 wood chisel
[chi/woo]
f ciseau à bois
e cincel de madera
i scalpello da legno
n houtbeitel
d Holzbeitel

2549 wood cutting tool; wood milling tool
[fre/woo]
f fraise à bois
e fresa de madera
i fresa di legno
n houtfrees
d Holzfräser

2550 wood drill
[dri/woo]
f foret à bois; mèche à bois
e barrena para madera; broca para madera
i trapano per legno
n houtboor
d Holzbohrer

* **wooden float → 2154**

2551 wood engraving chisel
[car/chi/woo]
f burin à contours de xylographie
e punzón de xilografía
i bulino da silografia
n houtsnijbeitel
d Konturenstichel; Linienstichel; Spitzstichel

2552 wood engraving contour knife
[woo]
f couteau à contours pour xylographie
e cuchillo de contornear para la xilografía
i coltello per incisione di legna; coltello per contorni da silografia
n contourmes
d Konturenmesser

2553 wood engraving gouge
[pla/woo]
f gouge ronde de xylographie
e gubia de xilografía
i sgorbia piatta da silografia
n houtsnijguts
d Rundeisen für Xylographie

2554 wood engraving hollow V tool; wood engraving parting tool
[chi/woo]
f gouge triangulaire pour la xylographie
e gubia triangular para la xilografía
i sgorbia triangolare per la silografia
n holbeitel voor houtsnijwerk
d Geissfuss

2555 wood engraving knife
[woo]
f couteau pour la xylographie
e cuchillo para la xilografía
i coltello per la silografia
n houtsnijmes
d Xylographiemesser

* **wood engraving parting tool → 2554**

2556 wood engraving scalper; wood engraving scauper; wood engraving scorper
[woo]
f biseau pour la xylographie
e escoplo para la xilografía
i sgorbia a lama piatta per la silografia
n platte houtsnijguts
d Xylographieflacheisen

2557 wood engraving scoop
[woo]
f gouge concave pour la xylographie
e gubia cóncava redonda para la xilografía
i sgorbia concava da silografia

n holle houtsnijguts
d Xylographiehohleisen

* **wood engraving scorper → 2556**

2558 wooden hammer
[ham]
f marteau en bois
e martillo de madera
i martello di legno
n houten hamer
d Holzhammer

2559 wooden nail; wood nail
[fas/woo]
f cheville
e clavo de madera
i caviglia; chiodo di legno
n houten pen
d Holznagel

* **wooden peg → 1743**

2560 wooden pin; wooden spike
[fas/woo]
f épite
e espiga de madera
i cavicchio
n deutel
d Holzdorn; Holzstift

2561 wooden plug
[fas/woo]
f bouchon en bois; tapon en bois
e taco de madera
i tappo di legno
n houten plug
d Holzdübel; Holzpfropfen; Holzstopfen; Holzstöpsel

2562 wooden shovel
[nut]
f pelle en bois
e pala de madera
i pala di legno
n houten schop
d Holzschaufel

* **wooden spike → 2560**

2563 wood file
[far]
f lime à bois
e lima de madera
i lima da legno
n houtvijl
d Holzfeile

* **woodman's saw → 2455**

* **wood milling tool → 2549**

* **wood nail → 2559**

2564 wood rasp
[far]
f râpe à bois
e escofina de madera; escofina de tabla
i raspa da legno
n houtrasp
d Holzraspel

2565 wood saws
[saw]
f scies à bois
e sierras para madera
i seghe da legno
n houtzagen
d Holzsägen

2566 wood screw
[fas/scr/woo]
f vis à bois
e tornillo para madera
i vite da legno
n houtschroef
d Holzschraube

2567 woodturner's finishing chisel
[chi/woo]
f ciseau pour repasser au tour
e cincel para repasar al torno
i scalpello per ripassare al tornio
n nadraaibeitel
d Nachdrehbeitel

2568 wood turning chisel
[chi/woo]
f ciseau de tourneur en bois
e cincel de tornero de madera
i scalpello di tornitore in legno
n houtdraaibeitel; houtslechtbeitel
d Drechslerbeitel

2569 wood turning gouge
[chi/woo]
f gouge de tourneur en bois
e gubia de tornero de madera
i sgorbia di tornitore in legno
n houtdraaiguts
d Drechslerhohlbeitel

2570 wood turning lathe
[lat/woo]
f tour à bois
e torno de madera
i tornio da legno
n houtdraaibank

d Drechselbank; Holzdrechselmaschine; Holzdrehbank

2571 **wood turning tools**
[gen/lat/woo]
f utensiles de tourneurs en bois
e herramientas de torneros de madera
i utensileria di tornitori in legno
n houtdraaiersgereedschap
d Drechslerwerkzeuge

* **workbench → 199**

2572 **working edge**
[car/joi/woo]
f bord d'usinage
e arista de trabajo
i bordo di lavorazione
n bewerkingskant
d Bearbeitungskante

2573 **working face**
[car/joi/woo]
f surface d'usinage
e superficie de trabajo
i superficie di lavorazione
n bewerkingsvlak
d Bearbeitungsfläche

* **working ga(u)ge → 1564**

* **working surface → 927**

* **working table → 199**

2574 **worn-out file**
[far]
f lime usée
e lima gastada
i lima consunta
n doffe vijl
d abgenutzte Feile

2575 **wrecking bar**
[chi/gen]
f pied de biche; pince de déballeur; pince monseigneur
e pie de cabra
i piede di porco
n sloopbeitel
d Geissfuss

* **wrecking bar → 425**

* **wrecking crane** *US* **→ 326**

2576 **wye**
[hyd/joi]
f union en Y
e unión en Y
i raccordo a 45°
n Y-verbinding
d Y-Verbindung

Français

Español

Italiano

Nederlands

Deutsch